How to Know a Crow

Candace Savage
Illustrated by Rachel Hudson

How to Know a Crow

The Biography of a Brainy Bird

DAVID SUZUKI INSTITUTE

GREYSTONE KIDS

GREYSTONE BOOKS • VANCOUVER/BERKELEY/LONDON

Contents

1. **Beginnings** 1
2. **Family Matters** 13
3. **A Kindness of Crows** 25
4. **Bird Brains** 37
5. **Party Animals** 49
6. **Life Choices** 61
7. **A Season of Dying** 73
8. **A Nest of Her Own** 85

Author's Note 96
Glossary 98
Resources 101
Index 102
About the Author and Illustrator 105

Beginnings

The clear light of early morning filters through the upper branches of the crows' nest tree. There's a chill in the breeze, a reminder of the winter just past, but the air is filled with birdsong and dappled with new-leaf green. Suddenly, it's glorious spring!

  Down below, people on foot and people in cars are moving this way and that. Seen from above, these groundlings look small and flat, as if you were peering down on them from an upstairs window. But raise your eyes—look up and out—and see how the view expands beyond the streets and over the rooftops to embrace a wide, encircling world of scattered trees and grassy fields.

  Large trees for nesting and roosting. Lawns and fields for foraging. This is a paradise for crows, exactly the kind of place they like best for raising their young.

You are about to enter the secret world of crows.

Until the 1950s, most American Crows lived on the edges of the forest and around farms.

Since then, more and more have moved into cities and towns, making them our *cawnstant cawmpanions*.

The crow family we are visiting are American Crows or, to use their official scientific name, *Corvus brachyrhynchos*. Although those fancy Latin words may look a little strange at first, they carry a lot of meaning. They are also fun to say if you enjoy tongue-twisting. The first word, **Corvus**, is pronounced **kor**-vus, and it refers to the whole group, or "genus," of crows and ravens everywhere in the world. (Members of the genus *Corvus*, together with close cousins like the magpies and jays, are sometimes referred to as "corvids.") The second word, **brachyrhynchos**, pronounced brack-ee-**rin**-kuhs, means "short-beaked" or "short-nosed." It indicates the particular kind, or species, of crow we are talking about. All told, there are about 45 different species of crows in various parts of the world, at home on every continent except South America and Antarctica, and in every country except New Zealand. All of them are black (or, occasionally, black and white), and all are quite big (compared to most songbirds, at least), with sturdy beaks, strong feet, and voices like car alarms. So no matter where you live or where you visit, there is likely to be some kind of crow kicking up a ruckus in the neighborhood.

On this spring morning, however, there is no cacophony of cawing in the crows' nest tree. But if you listen closely, you might hear a quiet rapping sound, as if a tiny hand were knocking on a tiny door. Tap, tap-tap, tap. The muffled noise is coming from inside the crows' nest—a big bundle of pencil-thin sticks, dozens and dozens of them, that have been woven together to form a large basket. As big as a kitchen sink, this heavy structure is supported by sturdy branches and wedged against the trunk of the tree.

Inside this fortress is a rounded hollow, at least as big as a cereal bowl. It is lined with soft material such as shredded bark, dried grass, tufts of animal fur, scraps of paper, and whatever else the adult crows have been able to gather. Occasionally, they will come across something strange and hilarious, like the pair of men's underpants that one crow used to line her nest.

Tap, crunch, tap-tap.

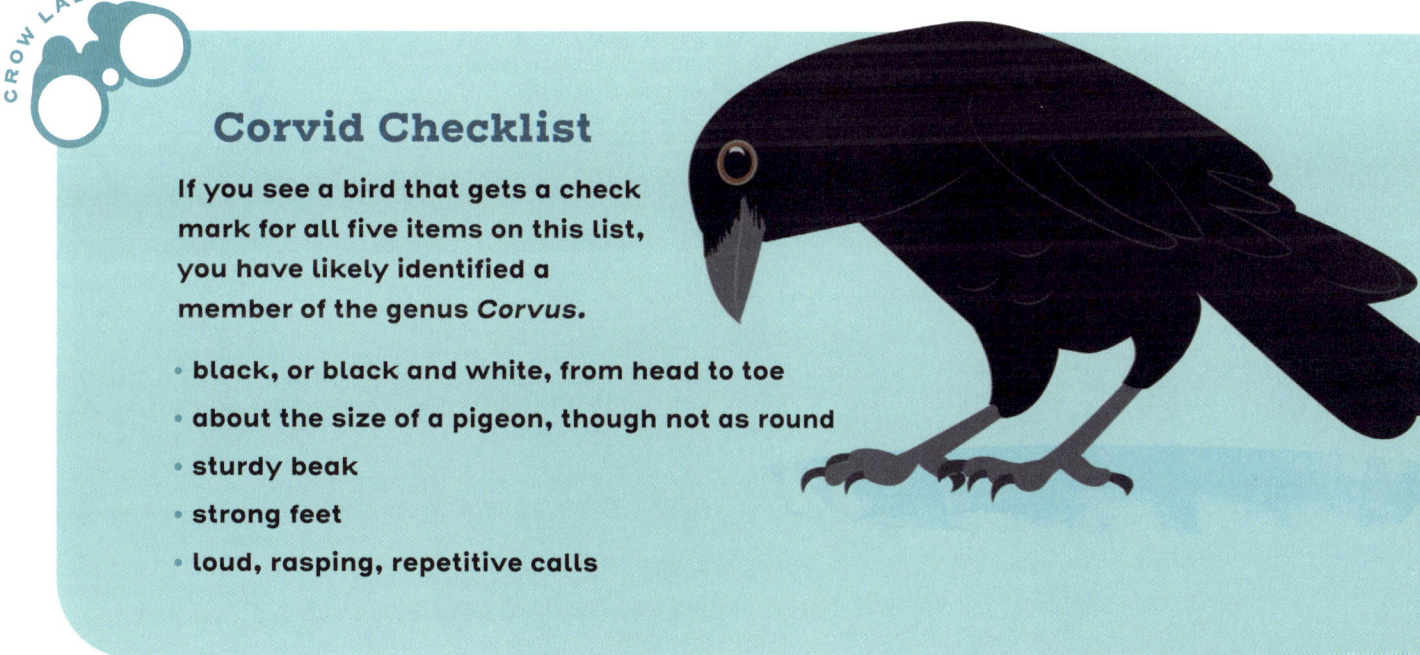

Corvid Checklist

If you see a bird that gets a check mark for all five items on this list, you have likely identified a member of the genus *Corvus*.

- black, or black and white, from head to toe
- about the size of a pigeon, though not as round
- sturdy beak
- strong feet
- loud, rasping, repetitive calls

Shh! Hatching in progress. Few birds are as quiet as nesting crows.

There are no undies in the bowl of this nest, just five sea-green eggs, each splattered with streaks and splotches of olive brown. The eggs narrow at one end and are a little smaller than a plum. You could cradle one carefully in your palm without much risk of dropping it.

The mother crow began laying three weeks ago, adding to her clutch at the rate of one egg each day. From the third day of laying onward, she kept her offspring safe by sitting closely over them, using the warmth of her body to nurture them during late-winter snowfalls and cold spring rains.

Now, however, she has roused herself from this drowsy vigil and is standing on the wide rim of the nest, staring down into the bowl. The quiet tapping is coming from one of the eggs—the largest of them, the one she laid first. Tap-tap, crunch, and suddenly, there it is! One of the eggs has pipped! Whatever, or whoever, is inside that egg has cracked a dent in it. With every tap, a triangular piece of loosened shell is pulsing up and down. Hours pass. Tap, tap-tap, tap.

Eventually, the shard of shell breaks loose and the tip of a pale beak pokes out. Working from the inside, the hatching crow continues to hammer at the shell, causing fragments around the opening to shatter and fall away. Like all baby birds, this one has an egg tooth, a sharp knob at the tip of the bill that helps break the shell. Even so, the little escape artist must stop every few seconds to rest and pant and catch his breath.

Finally, the top of the shell breaks loose and hinges out of the way. But even with this doorway open, the hatchling still has to fight for freedom, squirming and writhing to separate himself from gooey membranes and clinging bits of debris. He flails his stubby wings wildly and thrashes his spindly legs. And then, with a final burst of effort, he rolls clear of the broken shell and sprawls across the floor of the nest. His neck drapes over the still-to-hatch egg of a brother or sister. His scrawny chest heaves with effort.

This little hero deserves a name. Let's call him Oki.

Oki means "hello" in the language of the Niitsitapi, the Blackfoot and Blackfeet Peoples. Hello, Oki. Hello, hello!

A World of Crows

American Crows
Corvus brachyrhynchos nest in all ten provinces of Canada and in the contiguous United States. Most of the crows that breed in Canada fly south into the States for the winter.

Fish Crows
Corvus ossifragus are slightly smaller than American Crows. They live in the eastern United States, mainly along coastlines, rivers, and wetlands. Listen for their nasal caws: *uhn-uhn*.

Common Ravens
Corvus corax are among the biggest and boldest of all crows, with a wingspan equal to that of a large hawk. At home in North America, Europe, Asia, and North Africa, they also have the biggest range of all crow species.

Hooded Crows
Corvus cornix are striking birds with silvery-gray bodies that contrast with their black wings and heads. They live in much of Europe and parts of Asia, and they breed as far north as Arctic Finland.

Carrion Crows
Corvus corone are found in parts of Asia, including Japan. They are close cousins of Hooded Crows but entirely black. In places where both Hooded and Carrion Crows live together, they sometimes interbreed.

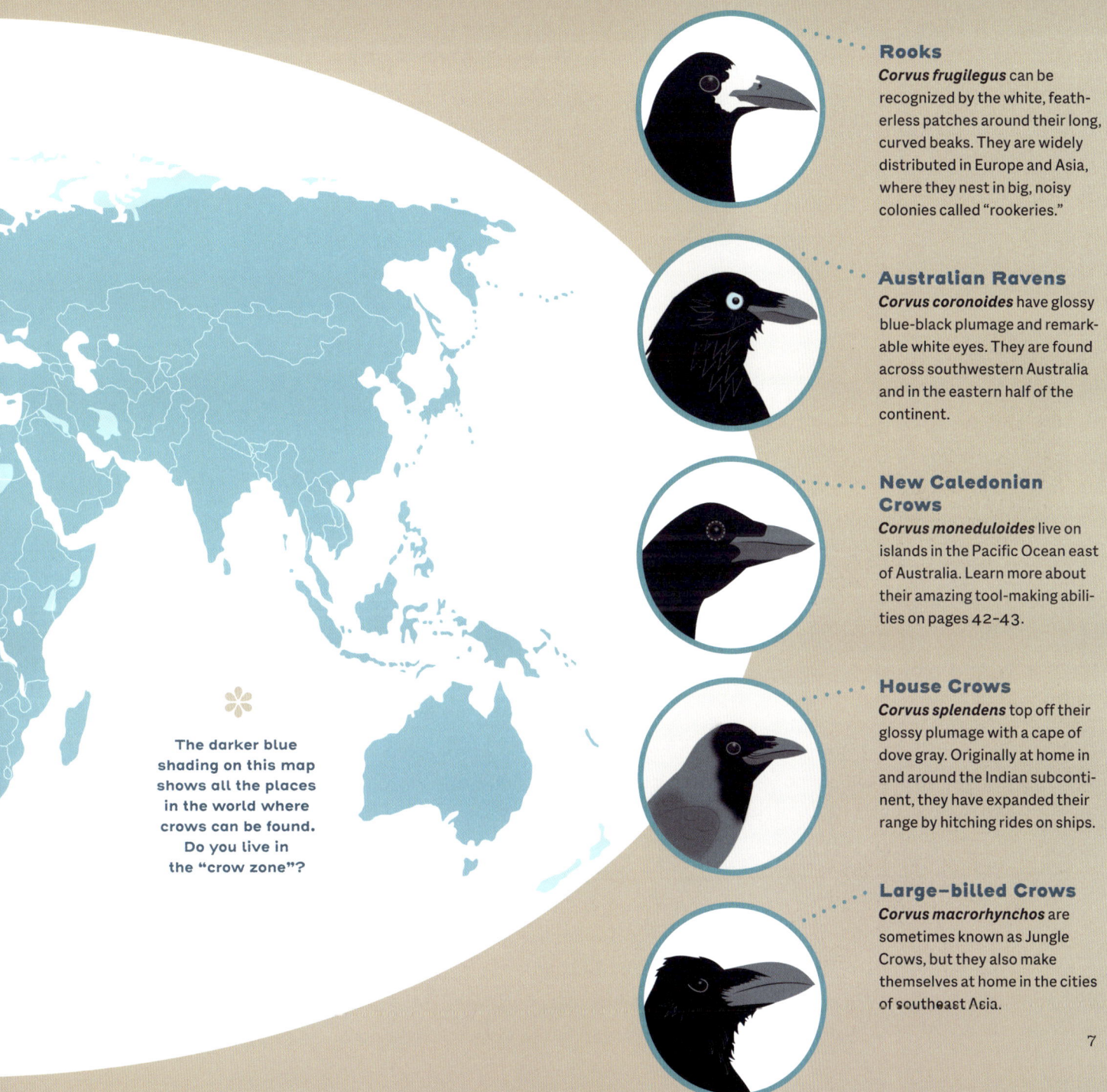

The darker blue shading on this map shows all the places in the world where crows can be found. Do you live in the "crow zone"?

Rooks
Corvus frugilegus can be recognized by the white, featherless patches around their long, curved beaks. They are widely distributed in Europe and Asia, where they nest in big, noisy colonies called "rookeries."

Australian Ravens
Corvus coronoides have glossy blue-black plumage and remarkable white eyes. They are found across southwestern Australia and in the eastern half of the continent.

New Caledonian Crows
Corvus moneduloides live on islands in the Pacific Ocean east of Australia. Learn more about their amazing tool-making abilities on pages 42-43.

House Crows
Corvus splendens top off their glossy plumage with a cape of dove gray. Originally at home in and around the Indian subcontinent, they have expanded their range by hitching rides on ships.

Large-billed Crows
Corvus macrorhynchos are sometimes known as Jungle Crows, but they also make themselves at home in the cities of southeast Asia.

Some birds, like ducks and chickens, are ready to run almost as soon as they pop out of the shell. Within a day, they are able to tag along behind their parents and pick up food for themselves. Scientists refer to these hatchlings as "precocial," meaning that they are developmentally advanced for their age. (If someone calls you "precocious," they mean much the same thing.) But newly hatched crows are different. They are "altricial," meaning that they require a lot of care from their parents and must be fed and kept warm. (Interestingly, the word "adolescent" comes from the same root, a quiet reminder that we humans are an altricial species too.)

American Crow

Common Raven

Which Crow Is Which?

So you've seen a crow—but which kind of crow did you see? With several dozen species to choose from, answering that question can be tricky.

The quickest way to narrow down the possibilities is to learn about the species that live near you. Check your public library or bookstore for a guide to your local birds. If you have a phone, you might want to download the free apps Merlin Bird ID and iNaturalist.

Then get outside and get happy. Bring a notebook with you and record the date and the species of crows you see. How many birds can you count? What are they doing?

CROW LAB

Oki, the new arrival in the treetop nest, is almost completely helpless. All head and belly, he weighs just half an ounce (15 grams), about the same as a couple of grapes. As an embryo growing inside the shell, he drew nourishment from the yolk of his egg, much as a human fetus takes nourishment from its mother's womb. At the time of hatching, the remaining yolk is drawn into the hatchling's body, leaving a rough patch on its belly where it was connected to the yolk sac. (The next time someone asks you if birds have belly buttons, you will know what to say. "Not exactly. They don't have innies or outies like us, but when they hatch, they do have umbilical scars where they were once attached to their yolks.")

Oki is pink and soft and nearly naked, tufted with feathery down. Too weak to stand or even to shiver, the hatchling has no way to keep himself warm. His bulging eyes are sealed shut, so he is as blind as a newborn kitten.

Just about the only thing Oki knows how to do is to open his mouth for food. His little body is a motion detector, and his ears—those slits on each side of his head—are alert for sounds. If anything jostles the nest, that big, red mouth snaps open, reaching up and quivering with eagerness. Just about any disturbance will set him off, whether it makes sense or not. It could be a gust of wind that shakes the tree, a nudge from another egg as it begins to hatch, or, with better luck, the plop of an adult crow returning with a juicy grub. Whatever the stimulus, the response is always the same. Open wide: *Feed me, feed me!*

Mammals   Reptiles   Birds

Do you think our hatchling crow is adorable, or does he strike you as a bit weird? Where would you rate him on a Cute-o-Meter? Birds are a familiar presence in our lives, and we feel close to them. We like to think of them as our "feathered friends." But scientists tell us that crows and other birds are only distantly related to us. If we picture the evolution of life as a magnificent and incredibly ancient tree, then birds (including corvids) and mammals (including us) appear on widely separated branches.

Those branches split off from one another more than 300 million years ago, at the time when the first vertebrates (animals with backbones) were crawling out of the oceans and swamps to live on dry land. Starting out as small lizard-like creatures, one line of animals went through a long series of strange and miraculous transformations to produce mammals like us. From the same starting point, the other evolutionary line, the one that leads to birds, went on to produce turtles, lizards, snakes—and dinosaurs.

Birds descend from theropods, a group of dinosaurs that includes celebrities like *Velociraptor* and the fearsome *Tyrannosaurus rex*, together with several smaller species that deserve to be just as famous. These small theropods could fly. Like modern birds, they had feathers, hollow bones, and scaly, four-toed feet. They laid eggs in nests, and some of them seem to have been caring parents. Recently, paleontologists working in southern China uncovered the fossilized remains of a feathered dinosaur that died while brooding a clutch of two dozen eggs.

People often say that the dinosaurs became extinct, but that isn't really true. A single group survived, and we call them "birds." No wonder our newly hatched crow looks astonishing and a bit odd. What else would you expect from the dinosaur in your backyard?

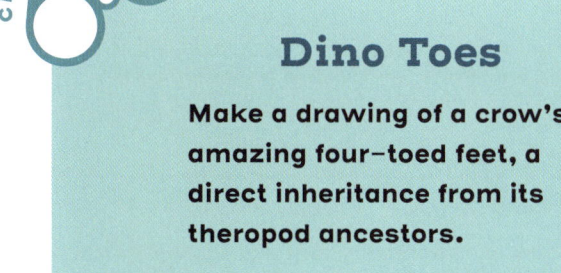

Dino Toes

Make a drawing of a crow's amazing four-toed feet, a direct inheritance from its theropod ancestors.

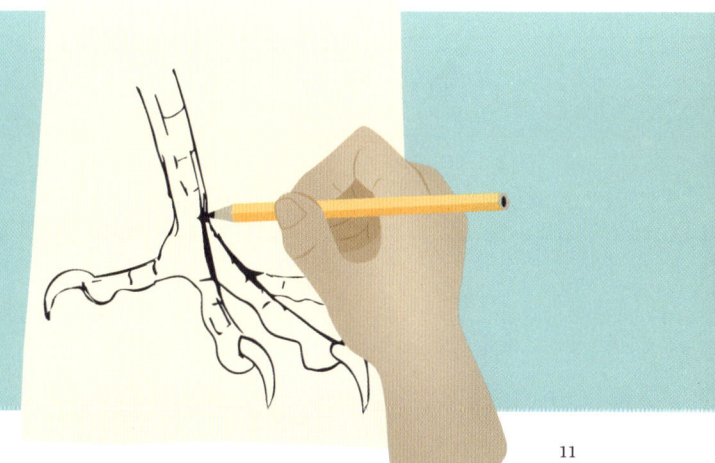

Family Matters

Six days have now passed, and up in the crows' tree, something exciting is going on. *Four* adult-sized crows are perched on the rim of the nest. Utterly quiet, they are all leaning forward and gazing into the interior bowl. At their feet is a tangle of sprawled limbs and bulbous bellies. Every one of the eggs has hatched! This is a good outcome for the crows, since eggs sometimes get jostled and addled (scrambled in the shell) or are infertile and fail to develop. But this time, five eggs have produced five nestlings. Their once-pink skin is darkening to a leathery slate gray, and the membranes that cover their eyes have opened just a crack.

The watchers at the nest look on in silence. The only sounds to be heard are the faint squeaks and yelps of the nestlings, desperate as always for food. With their feet spread out behind them as anchors and their wing tips as supports, they stretch their stringy necks to the limit. Their red-lined mouths bob and wave, like a bouquet of ravenous flowers. Oki is still the biggest, though the others are catching up.

Rock-a-bye baby
in the treetop,
When the wind blows,
the crow nest will rock,
When the bough shakes,
mouths open wide,
Begging for someone
to put food inside.

❋

When species that are not closely related face similar environmental challenges, they sometimes change to become more like one another.

Scientists call this process "convergent evolution."

After a few moments, one of the watchers tips its head deep into the bowl and thrusts its sharp bill toward the outstretched mouths. Its throat flexes as it coughs up a saliva-covered glob of earthworms from its gular pouch and places it into that eager gullet. With a quick tilt of the head, the caregiver checks what it has just done. Perhaps a little too much? It deftly retrieves a blob of the food and drops it into the next mouth in line. Any false movements with its heavy beak could cause injury or even death to the nestlings, but no harm is done. The crow's movements are precise and gentle, exquisitely careful. We are only distantly related to these dino-birds, but there is something about this gesture that seems humanlike and familiar.

Hide and Go Seek

A crow nest is a massive bundle of sticks stuck up in a tree, so you might think it would be easy to find. Not so fast, Sherlock. This is a task that calls for a super-sleuth. Follow the clues.

- **Check the date.** Crows nest early in the season, as winter tips into spring.

- **Watch for crows carrying sticks.** Try to see where they take them. If they head high up into a large tree, you might be onto something.

- **Look for a scattering of twigs on the sidewalk or under a tree.** Crows sometimes drop sticks when they are gathering materials or building their nest.

- **Listen for loud, insistent, whining calls.** That's a female crow who has started incubating eggs. She's reminding her mate and helpers, again and again, that she needs to be fed.

- **Be kind. Be considerate. Be warned.** Do not pursue or harass crows, and do NOT climb to their nests. If they decide you are a threat, they may dive and shout at you! If you do happen to annoy them—or if they mistake you for someone they dislike—you will have to avoid the area around the nest for a while or carry an umbrella as a shield.

> **PRO TIP**
> If you want to figure out where a nesting crow is going, look at it out of the corner of your eye or pretend to be looking at something else. If you gaze at a crow directly, it will become even more wary than usual and will not go to its nest.

Did you think it was strange to see so many birds crowded around the nest? Then you have noticed something important. Believe it or not, crows are songbirds, members of the same evolutionary lineage as tunesmiths like robins and larks. For most members of this group of birds, raising a nest full of young is a mom-and-pop operation. The male and female parents work together to feed and defend their brood. This is often the pattern for American Crows as well. A mated pair of crows are perfectly capable of rearing young on their own, and many of them do.

But often the parents have help—or what looks like help, at least. In most cases, the additions to the workforce are the pair's own offspring from previous years. In other words, they are the older brothers and sisters of the nestlings that they are helping to look after. Other relatives (aunts or uncles, for example) and "adoptees" from other families may also be part of the group. Scientists refer to this kind of team effort as "cooperative breeding." It is found only in a small fraction of bird species, making it special and rare.

Some of the extra helpers, or auxiliaries, just seem to hang about, but others take an active interest in everything that is going on. Is it time to gather sticks? Construct the nest? Watch for danger? These ready-eddies are on the job. But it's a bit like getting your little brother or sister to help you clean up your room. "Helpers" don't always seem to be very helpful. Scientists have not been able to figure out who benefits most from these family relationships. Maybe the real winners are the young birds who get to practice their skills in the comfort and security of their home territory. Whatever the pluses and minuses, this breeding system must have advantages, because it is what American Crows tend to do. For them, raising a nest full of hatchlings is a family affair.

Back in the 1940s, when scientists first noticed extended families of crows helping at nests, they could hardly believe their eyes.

Fewer than 8 percent of bird species are cooperative breeders.

Counting Crows

A crow sitting all by itself, cawing into thin air, is waiting for an answer from nearby friends or neighbors. Crows live in flocks and families, and are rarely truly alone. Look and listen. How many crows do you see? How many can you hear? What can you tell about their relationships? Are they friends or foes? What messages do you think they might be communicating to each other?

Even with extra caregivers on the job, things can still go wrong. A storm can come up without warning and send the nest, and its precious contents, crashing to the ground. Some of the nestlings might get sick. And if food is scarce that year, some (especially the last to hatch, often the smallest) could even starve. Through many weeks of preparation, the parents and their helpers have taken steps that increase the nestlings' chances of success.

In late winter—before the eggs were laid—the family had already started defending their territory against other crows. You might have heard them shouting at intruders with rapid-fire caws or watched their darting flight as they chased off unwelcome visitors. This is the crows' way of putting up No Trespassing signs around their territory and claiming the food and other resources for themselves. Crow territories vary widely in size, from as small as a square block to as large as a square mile, depending on how much food the area provides.

Then, as winter turned to spring, the parents invested time and care in selecting the best possible site for their nest. Crow parents often go to the trouble of laying a few test-sticks in one spot and then another before making their choice. It takes them a week or even two to set up their nursery, tinkering with twigs to build a structure that is strong and cozy.

The whole family helps build the nest, but the mother crow arranges the soft lining in the bowl.

After all, she is the one who will sit in it for days and weeks to come!

Crafty as a Crow?

Challenge yourself to build a nest out of natural materials like sticks, grass, and mud. How big a pile of twigs can you gather? How will you hold them together and secure them in place? What other decisions will you have to make? Would your nest weather a storm? Imagine doing all this construction and problem-solving using only a beak!

Once the eggs are laid, the mother crow incubates them, night and day, for around three weeks, allowing herself only a few minutes off at a time for stretch and bathroom breaks. Even after the eggs have hatched, she continues to spend long hours preening her new offspring, cleaning up after them, and sitting on their squirmy bodies to keep them warm.

Through the long days while the mother is sitting, the father and the helpers feed her at the nest and stand guard in nearby trees, quiet, alert, on the lookout for trouble.

Trouble can come in the form of an egg-eating squirrel, a prowling raccoon, or—most alarming of all—the whispered flight and yellow glare of a Great Horned Owl. With its ripping beak and crushing talons, this powerful predator can tear the heads off adult crows and wipe out entire nests. (If you ever find the severed head of a bird lying on the ground, you will know that a Great Horned Owl has been doing what it does best.) If crows hate and fear anything, they hate and fear Great Horned Owls. The mere sight of an owl dozing in a tree ignites a frantic outburst of harsh cawing that attracts the whole family, even drawing the mother off the nest. For the moment, No Trespassing signs are forgotten, and crows come streaming in from every direction to mob their enemy. They congregate around it, screaming, screaming, screaming. Roused from its daytime slumber, the owl eventually gives in and speeds away, with a cloud of black avengers on its tail.

No Great Horned Owl will menace Oki's family. No storm will blow their nest down. But one afternoon, when the mother crow has slipped away from the nest for a moment, a dark shape darts through the branches, talons at the ready, and snatches one of the baby crows. The parents and helpers erupt into a crazy commotion of cawing; they chase and dive—but nothing more can be done. The Cooper's Hawk speeds off to feed its own brood, and the crows now have four young. Oki is one of the lucky ones.

The next time trouble comes to the nest, it will be carrying a ladder.

Caw patrol,
to the lookout!
No job is too big,
no crow
is too small.

Welcome to the Crows' Nest

High-Rise Living at Its Best

Health and Safety Standards

- The nests are built on a platform of strong branches and are typically close to the main trunk for extra protection and strength.

- The foundation of the nest is constructed of hundreds of twigs, mostly smaller around than your finger and shorter than your arm. Crows work the twigs into place with the broad ends out, creating a strong fortress.

- Crows often cement the twigs together with a floor of mud and wet grass.

- The inner bowl is lined with soft materials like dried grass, shredded bark, moss, flowers, hair, paper, twine—whatever is handy. (Undies optional!)

Four-Star Accommodations

- The ideal place for a crow nest is in the top third of a large tree, between 30 and 100 feet (10 and 30 m) above ground. That's about the height of an apartment between the third and tenth floors.

- The best sites provide dense cover on all sides, for protection from the spying eyes of predators. Large evergreen and broad-leaved trees both do the job well.

The Main Attraction

- Although a female crow can lay between three and seven eggs, the average is four or five. Only the luckiest few hatchlings survive the perils of their first year of life.

Built to Last

▽ Nests often remain visible and apparently in good shape for several years, but crows seldom reuse them. Abandoned crow nests are often occupied by squirrels or used as nesting platforms by hawks and owls. Raccoons also find them a cozy place to nap.

No-Star Option

▽ If there aren't any trees or shrubs available, crows will nest on the ground, but this is risky. Ground nests are vulnerable to cats, dogs, and other predators.

A Kindness of Crows

Before you know it, another couple of weeks have sped by, and Oki and his three surviving siblings are now around 20 days old. Late spring has blossomed into summer, and the nestlings are undergoing an amazing makeover.

    Instead of the straggly, sprawling bodies we saw the last time we looked into the nest, there are now four beefy black birds, almost as hefty as pigeons, squished in wing to tail and beak to breast. These guys have been fed to the brim! Day after day, the father crow and the helpers have been on the run, shuttling from the nest to the field and back again, delivering menu items that match the age and stage of the nestlings. With the youngsters now able to keep themselves warm, the mother has finally been freed from round-the-clock nest duty, so she is available to fetch and carry as well. Beetles, caterpillars, grasshoppers, and grubs; spiders, frogs, crayfish, and mice; french fries, junk food, and bits of dead things: it has all gone into those demanding gullets.

In less than 20 days, nestling crows balloon from the weight of a couple of grapes to the weight of a grapefruit!

Of course, what goes in must come out, and the nestlings produce lots of poop. Instead of typical adult-bird splats, they eject their waste in tidy little packages called "fecal sacs." The caregivers pick up these sacs and either eat them (when the nestlings are very young) or drop them well away from the nest (when the nestlings are a little older and there is less food value in their excrement). When the nestlings are old enough, they back up to the edge of the nest, waggle their tails, and bombs away!

Very occasionally, the nestlings are fed the eggs or nestlings of other birds, a fact that makes some people upset. These folks believe that crows are wiping out our favorite songbirds, a situation that would be alarming if it were really happening. Scientists tell us that, yes, many species of songbirds are becoming worryingly rare—but that crows are not to blame. Eggs and nestlings typically make up only a tiny portion of the food that is fed to baby crows, around 1 or 2 percent of their diet. At this modest rate, small songbirds typically produce enough "extra" young to be able to cope with the losses. American Robins, for example, can lay up to three broods, for a total of fifteen eggs, in a single summer. To keep the population steady, each pair only has to produce two surviving offspring in their lifetimes. So there's a lot of surplus in the system—a lot of room for give and take—and the occasional snatch-and-grab by a crow (or a Cooper's Hawk, for that matter) is not a calamity. Domestic cats, by contrast, can make a meal of local songbird numbers, so it is important and necessary to keep Puss-Puss indoors, especially during nesting season.

The main reasons for declining bird populations come as no surprise: loss of quality places to live and the climate crisis.

CROW LAB

Watchful Bird-Watching

Let's say you find a sweet little nest of songbirds' eggs or young. As you ooh and aah over this discovery, look around to see who might be watching or listening. If a crow or a jay notices what you are doing—or hears the alarm calls of the parents you have disturbed—it will likely wait until you are gone and then drop by for a tasty snack.

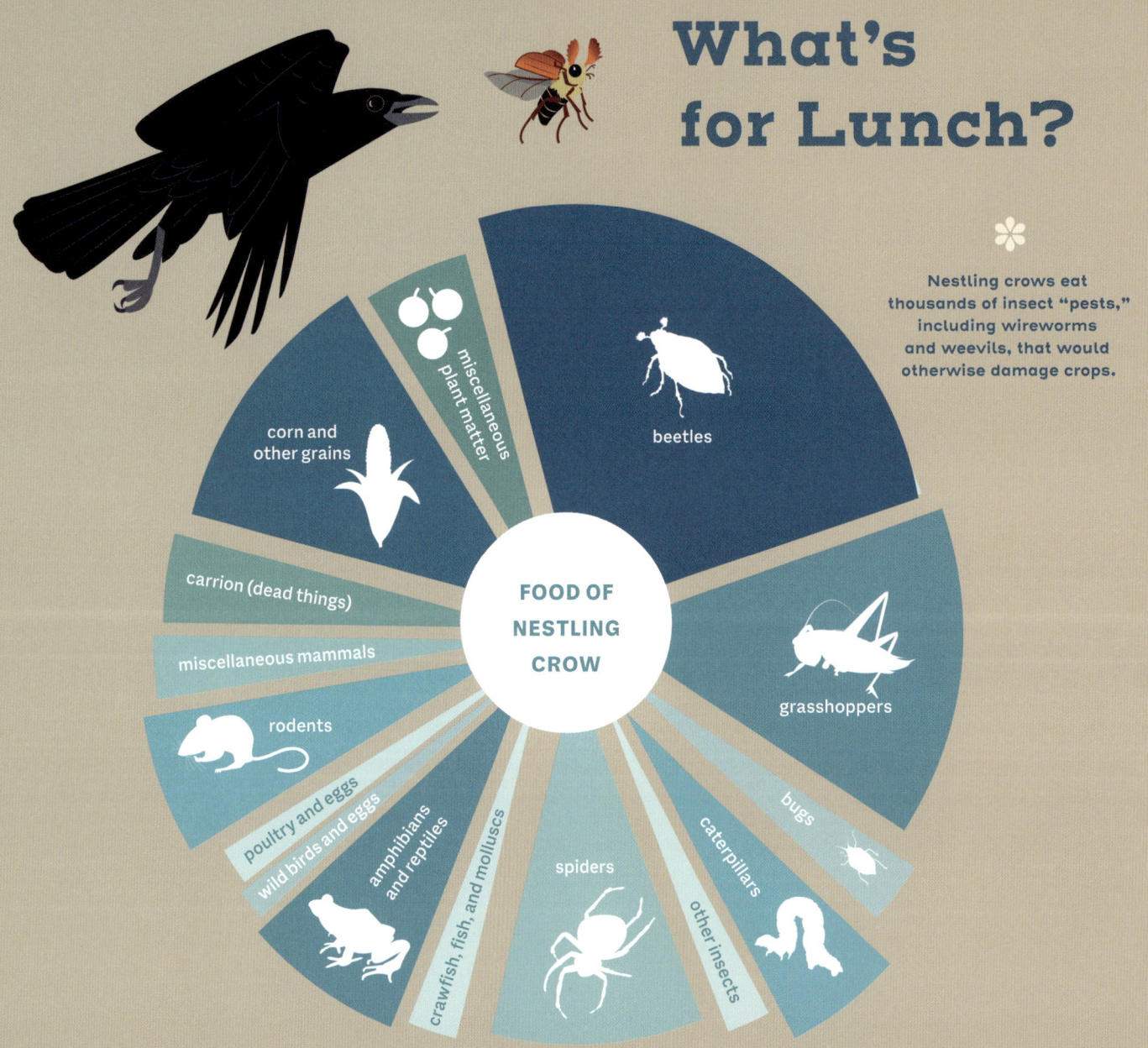

Like most baby songbirds, nestling crows eat a *lot* of insects. This chart shows the proportions of different kinds of foods fed to nestlings in farming areas. Baby crows that live in town receive a similar diet, plus a side order of compost, fast-food scraps, and other human trash.

Nestlings that grow up in cities tend to be a little smaller than their country cousins. Too much junk food is bad for crows too.

Day by day, Oki and his nestmates have been getting cuter and cuter. No longer reptilian and bony, their bodies are smoothed and rounded with a soft layer of contour feathers. Quill-like sheaths, or pinfeathers, have pushed through the skin along the outer edges of their wings. Inside them, flight feathers are forming. So itchy! The nestlings twist and wriggle, working the sheaths with their bills. If the weather turns cool, their mother still folds the youngsters under her body. When it's very hot, she steps back, and they drape themselves over the edge of the nest, open-mouthed, like so many wilted flowers. Sometimes, the mother or one of the helpers stands above them with wings outstretched, shielding them from the sun. But not even these doting caregivers can protect them from what is about to happen.

Caw-caw-caw! Danger! Someone is sounding the alarm. Oki and the others know almost nothing about the world beyond their nest. Even with their blue-gray eyes now fully open, their range of vision barely extends beyond the tips of their bills. At the grand old age of 25 days, they are still literally babes in the woods. All they know is what is familiar—the thump of a caregiver landing on the nest, the voices of their family members, the rustle of leaves in the wind. But the sounds they are hearing now are new and unnerving.

The murmur of human voices. The scrape of a ladder being maneuvered into place. The crack of a broken branch.

Caw-caw-caw-caw!

The dictionary word for a group of crows is a "murder." That's so mean! Around their nests, crows are downright cuddly.

Oki is up and alert, scrambling to escape—a dangerous move for one still so young and inexperienced. But then darkness falls as something that doesn't quite feel like mother suddenly covers the nest. A hand reaches under the covering, grasps the nestlings one by one, places them in a bucket lined with soft cloths, and snaps the lid shut. Down on the ground, a team of scientists weigh and measure the "kids," as they call them, draw samples of blood, and fit them with identifying bands and tags. Oki is now white-red-green, USGS 9875-54321; the tags on his—make that *her*—wings read MX. A blood test tells us that she is a female. It is all quick and painless (the youngsters do not squawk or wince), but the older members of the family are clearly unhappy about what is going on. As the banders return the kids to the nest and pack up to leave, the crows wheel and shout overhead, giving the intruders a noisy send-off.

In general, people are really terrible at telling individual crows apart. That's why the researchers have gone to the trouble of marking each nestling with a unique combination of bands and tags. Crows, by contrast, have no trouble identifying people as individuals and distinguishing friends from foes. In one famous experiment, researchers disguised themselves with caveman masks when they climbed to nests. Later, unmasked, they could walk freely around the area without triggering a response. But the second a scary, baby-handling caveman appeared in the crowd, mobs of crows rallied around to dive and shout. The attack force included not only birds who had seen the masked monster in action but also their neighbors and, in time, their offspring.

Crows are not just good at recognizing people's faces; they can often also read our intentions. If known individuals walk through the neighborhood, they may be allowed to pass without a fuss. But if those same people show up with a ladder and nets, off goes the crow alarm. To keep the peace, researchers often keep their pockets stuffed with treats to toss to the crows they meet, hoping to make some friends with random acts of peanuts.

❋

A "culture" is a collection of beliefs and traditions that pass from generation to generation through learning.

Scientists think that the shared knowledge of crows is a kind of culture.

May I Please See Your ID?

Recognizing crows as individuals is not easy for mere humans to do. But by looking closely, it is sometimes possible to make interesting distinctions. Even if you are not sure exactly who's who, you can keep notes or make sketches of what you observe. Here are a few differences to look for.

> **PRO TIP**
> Now and then, crows have personal quirks that make it possible to pick them out of the crowd, like an unexpected blob of white on a wing, a limp, a favorite perch or vocalization, or a feather that's always askew. Snap a photo! Take notes!

1. Fledglings, just out of the nest
EYES: blue fading to gray
INSIDE OF MOUTH: bright pink
PLUMAGE: brownish-black, not glossy, scruffy-looking

2. One-year-olds
EYES: brown
INSIDE OF MOUTH: pink, though beginning to darken
PLUMAGE: Glossy, but wing and tail feathers bleached and brownish. Ends of tail feathers pointed (not squared off). When the bird is perched, the tail feathers form an upside-down heart.

3. Adults
EYES: brown
INSIDE OF MOUTH: black by age two
PLUMAGE: Glossy black with a shimmer of iridescence. Ends of tail feathers squared off, forming a smooth fan when spread.
OTHER: males *slightly* larger than females

Nesting females have a brood patch—a bare, blood-warmed area on the breast for keeping eggs and hatchlings warm. If you ever get to see this patch, give yourself a gold star. Since the females spend almost all their time on the nest during breeding season, few people ever get to see it.

Join the Bird-Banding Band!

Every year, researchers around the world band *millions* of birds in the hopes of learning more about how they live, where they travel, and, sometimes, how they die. This is a big effort, and it only pays off if members of the public—people like you and me—notice the bands and report the Who, What, When, and Where of their sightings. To participate in this important scientific work, look up "How to report a banded bird" on a computer at home or in a library and click through the steps. As a thank-you for reporting, you will receive an official Certificate of Appreciation.

Back in the nest, life quickly returns to normal. Eat, poop, sleep, stretch, fidget, flap: repeat. The nestlings take no notice of the bands and plastic "feathers" that bedazzle their feet and wings. But thanks to these identifiers, we (and the researchers who did the banding) will be able to follow the young crows as they take their next big step, out of the nursery and into the wide world. This is going to be an adventure!

Bird Brains

Compared to people, songbirds grow up fast. A five-week-old human infant can't do much more than she or he could at birth. At that age, by contrast, a robin or a wren is well on the way to caring for itself, and its parents are busy producing a second, or even a third, brood of nestlings. But crows take their time. Because their nestlings grow up relatively slowly, the parents can only fit one brood into the short span of summertime.

By the time Oki is five weeks old, she is almost as large as an adult, but she is far from grown up. She and her siblings are like supersized toddlers, all rumpled and untidy, with comically short tails and floppy, unmanageable wings. Alone, in pairs—all together—they hop onto the rim of the nest, hold on with their toes, and beat the air like mad. Unable to fly, they have nonetheless started venturing outside the nest, hopping along nearby branches, going a little farther each time, and then hurrying back home.

But the nest is no longer the sanctuary that it used to be. By now, all the crow predators in the neighborhood have had time to map its location and observe what it contains: a nice big juicy bowl of nestlings. It is time to get moving!

Imagine if you'd had to look after yourself when you were five weeks old!

A crow's brain is smaller than a walnut, but it is packed with 1,500,000,000 neurons (as many as some monkeys and five times as many as pigeons have).

That's a lot of bright ideas.

As the eldest, Oki leads the way, hopping and fluttering from branch to branch, twiddling with twigs as she goes, and pausing to cry for food. One day, one of her caregivers sweeps past and deposits a glob of coughed-up insects a little farther out on the branch, as if trying to lure her along. *Caw-caw*, the caregiver shouts. *Come and get it.*

Really?! Did that caregiving crow actually make a plan and purposefully try to lure Oki away from the nest? Are these dino-birds bright enough to do things like that? And come to think of it, why do crows mature more slowly than most other songbirds do? Scientists think that the answers to these questions are likely related.

One reason that nestling crows need time to develop is that they are not just growing bigger—they are also growing smart. Obviously, a crow's brain isn't very large: it weighs less than a shelled walnut. (Imagine how hard it would be for a bird to fly if it had a massive head!) Yet compared to the weight of their bodies, a crow's brain is heavier than you would expect. The brain-to-body ratio of crows is higher than that of most other animals and is close to that of chimpanzees, our own closest relatives. The only other birds that come near this mark are those well-known smarty-pants, the parrots.

But size isn't all that counts. Not surprisingly, dinosaur brains are put together quite differently from our own. Unlike mammals, birds lack a cerebral cortex, the wrinkled, multilayered structure at the front of our brains that lets us think, create, learn, and generally have a good time. (It's what you and I are using right now to think about bird brains.) Until very recently, this absence was interpreted as proof that birds are "dumb clucks," mindless robots that go through their lives on autopilot. (*See an open mouth. Plop food into it.*) The assumption was that if birds didn't closely resemble us, they couldn't possibly be intelligent.

Well, that wasn't very smart.

It turns out that there is a lot going on inside those feathered heads, more than anyone ever suspected. Like us, birds also have structures in the forebrain that they use to think, create, and learn. Where we have a cerebral cortex, they have a structure called (wait for it!) a "nidopallium caudolaterale" [nih-doe-**pa**-lee-um caw-doe-**la**-ter-a-lay], or NCL, for short. Not only do crows have surprisingly large brains, they have surprisingly large forebrains too.

And there's more! The forebrain is made up of nerve cells, or neurons, that are wired together in intricate ways. The more neurons there are in these circuits, the more connections (synapses) the brain can make. Since birds have smaller neurons than mammals, a crow can pack a lot of nerve cells into its nut-sized brain, about the same number as a much-larger monkey. Add it all up—exceptionally large brain + exceptionally large forebrain + exceptionally large number of synaptic sparks—and you've got yourself a lot of firepower.

Growing all that mental capacity takes time, and that's one reason Oki and her siblings have spent so long in the nest. Now she has to make sense of the world and her own intelligence.

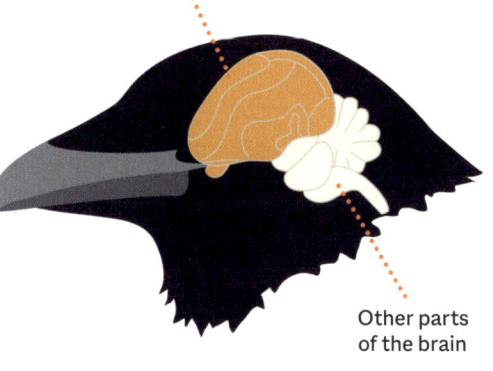

Brain structures responsible for complex behaviour

Other parts of the brain

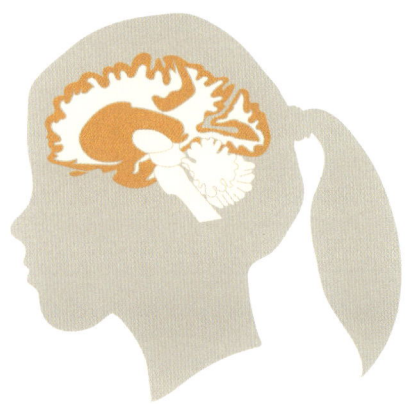

When a bird is sleeping, half of its brain rests while the other half remains awake and alert. Some birds use this handy trick when they are migrating.

Day by day, Oki has been getting bolder, venturing farther from the nest and staying away longer at a time. And then one day, when she is just over five weeks old, it happens. She leaves and doesn't come back. As evening falls, she tucks herself into the cover of dense branches, snugs her beak into her breast feathers, and falls asleep. Although she is by herself for the first time in her life, she is not truly alone. Around her, she can hear the comforting rustle of her caregivers and siblings settling down nearby. Within days, the nest will be abandoned, no longer of any interest or use to the crows. The adults will start construction afresh next season.

Just like that, Oki has fledged, or left the nest. Smart as she is, she didn't need to assess the situation and decide it was time to take the leap. Like a human toddler taking his or her first steps, she is just doing what comes naturally. The impulse to leave is encoded in her genes and is as much a part of her crow-ness as her black plumage.

Like a human baby who can't help putting things in his or her mouth, Oki has an irresistible urge to peck at objects and pick them up with her beak. Is this pebble edible? What about this dry leaf or this moldy hamburger bun? If it isn't good to eat, can I turn it into a toy and play drop-and-catch? Powered by curiosity, her marvelous brain lights up with every new experience.

String Games

If you are lucky enough to have a family of crows who regularly visit your yard, here is an experiment you can try. You will need a length of string (about as long as your arm), a piece of meat (half a wiener, for example), and a tree with a strong branch.

Tie the meat to the string. Tie the other end of the string to the branch, so that the meat is dangling down. Then hide out of sight and wait to see what happens. Do the crows notice this strange source of food? Can they figure out how to get hold of the prize?

In experiments with captive ravens, some birds quickly solved the problem by standing above the suspended food and using their beaks to pull the string up, length after length. Between tugs, they held the string in place with their feet. It was clear that they understood the problem they were solving. Wild ravens have been known to use the same technique to pull up lines set for ice fishing.

From Toys to Tools

Young crows love to fool around with stuff. They pick up sticks and put them down. They play drop-and-catch with pine cones. They bury bits of broken peanut shell and then dig them up.

Crows also occasionally use objects as tools. A pine cone can be deployed as a missile to bomb an intruder. Bam! Direct hit! A wedge of bark can be sharpened to probe for insects lodged in a crevice. A busy lane of traffic can be repurposed as a nutcracker. In this scenario, a crow flies down, drops a hard-to-crack nut in the path of oncoming cars, and retires to a lamppost to watch. He or she waits until the job is done, and the traffic lanes are clear, before coming down to claim the meal. Different crow species in North America, Europe, and Asia have all come up with this same high-tech solution.

But the Gold Medal for Exceptional Achievement in the Use and Manufacture of Tools goes to a species of crow that lives on a cluster of islands in the South Pacific Ocean. The New Caledonian Crow uses its specialized beak to shape two different kinds of implements. One is a twig with a hook nibbled at one end, while the other (made from the spiny-edged leaves of the pandanus tree) resembles a sturdy, serrated blade. Both are used as probes for extracting large beetle larvae from holes and crevices in trees. People used to think that humans were the only species clever enough to make tools, but we have had to make room on our pedestal for these brainy birds.

FACING PAGE: Provided with a straight piece of wire, this captive New Caledonian Crow quickly shapes it into a hook and uses it to score a hard-to-reach treat.

Oki has so much to learn. At first, her wings seem to have minds of their own and take her in unplanned directions. Sometimes, she has to recalculate her flight plan in midair or bob around in an undignified way when the branch she is trying to land on turns out to be nearer or farther away or springier than she expected. One day, her father approaches and calls to her, then flies directly off. She flutters into the air but ends up going the wrong way and plops down in a bush, dangerously close to the ground. Back her father comes and repeats his actions, flying away in the same direction as before. Is he trying to lead her to a safer spot? After several more tries with this back-and-forth, she finally gets herself organized and tags along after him.

CROW LAB

Crow Down. What Now?

It isn't unusual for a fledgling crow to end up on the ground. It might have fallen out of the nest or taken fright and jumped before it was really ready to leave. Or maybe it fledged normally but then descended by degrees, sinking a little lower every time it tried to use its wings. Now, no matter how hard it flutters, it can't get itself off the ground. So that's where you find it, hiding under a bush. It looks so cute and defenseless and lonely that you want to take it home and look after it.

Here's a word of advice: DON'T. Unless you are a licensed wildlife rehabilitation expert, taking a wild bird into your care is against the law. Besides, that little bird's parents are almost certainly nearby and are still on the job. It will fare much better in their care than it would in yours.

If the fledgling is at risk of being grabbed by a cat or dog, you could ask an adult to catch it and place it up in a shrub or tree. As long as the bird seems healthy, the best thing to do is usually nothing at all.

If the fledgling looks sick or is injured, call your local wildlife rehabilitation center for help. While you have them on the phone, ask if they need volunteers. Maybe you can sign up to help other sick or injured animals.

PRO TIP
Birds have sharp beaks and claws. If you recruit an adult to help, remind him or her to wear gloves.

Oki still relies on her caregivers for protection and support. When they shout out a warning, she rushes for cover. When she is hungry, she cries to them for food. If you hear a nasal *waa-waa-waa* coming from a shrub, it's likely the call of a fledgling crow hoping for lunch. Although she is now able to find some food for herself—that moldy bun hit the spot—there is a limit to what she can learn through trial and error.

Many of the crows' staple foods are tucked away out of sight. Grubs are buried in the dirt; kernels of corn are wrapped in husks; scraps of pizza are bundled into trash cans. Oki watches as the older crows uncover hidden treasures and then does her best to imitate their actions. If an adult digs in the ground, she rushes over to join in, sending fallen leaves and bits of debris flying in every direction. Then—what's this? A wriggly centipede goes down the hatch.

One day, she is standing beside one of her caregivers—head cocked, watching his every move—as he smashes a peanut shell open with his bill. When the treat inside is revealed, he could easily gobble it up.

Instead, he steps back and lets Oki have it. It is the crow version of Lunch and Learn.

We cannot know what Oki and the other members of her family are thinking or feeling when they do these intriguing things. They are dinosaurs, after all—strange and amazing beings. But scientists studying crows in a laboratory have watched the birds' forebrains light up when they notice what is happening around them, consider what they know, and decide what to do. The world of a crow is alight with awareness.

Ode to a Crow

As fledglings, crows are bold and bumbling and sometimes make us laugh. They are an ideal subject for a form of humorous verse called a limerick. Here is an example to get you started. Notice the *dumpety dumpety* rhythm and the AABBA rhyme scheme. Make yours sunny, punny, and funny! Write on!

(A) There once was a young crow who fledged
(A) And landed high up on a ledge.
(B) Someone brought him some food
(B) Which brightened his mood,
(A) But then he got stuck in a hedge.

Party Animals

Big, black birds of a feather flock together.

Summer ripens into early fall, and Oki is now four months old. No longer a "toddler," she has developed into a zippy, zestful flier and spends her days zooming around the countryside with her family. Evening finds her high up in the leafy canopy of a large deciduous tree, turning a now-brown eye to catch a view of her surroundings. Down below—beyond the streets and over the rooftops—lies an encircling vista of scattered trees and grassy fields, with everything a crow could need or want. The whole wild world is beckoning to her.

Now that Oki and her siblings are able to fly, she and her family have started leaving their territory to go on outings. Sometimes, they stay away overnight, for a kind of sleepover. Other times, they are out and about for several days before returning home. These adventures usually begin in the early evening, before sundown. Someone starts to caw, and suddenly, there's a rush of wings and the whole family rises into the air. Off they streak, one after another, flying purposefully in the same direction.

Play is a way of learning. It is something smart animals do. When someone tells you to stop fooling around, you might want to mention this!

Their destination is a favorite grove of large trees, or a roost, a mile or two away. Here they are joined by half a dozen other families, each of which includes parents, other group members (if there are any), and young of the year like Oki. The total community may number two or three dozen crows.

In the fading light before bedtime, the kids from all the family groups fool around together on the ground, pecking and poking at whatever they can find and, sometimes, at one another. If someone picks up an old bone, someone else is likely to grab hold of an end for a quick game of tug-of-war. There's a constant flurry of activity: jumping, kicking, squabbling, squawking. But when darkness falls, everything goes still as the entire flock of big black birds disappears into the larger blackness of the tree canopy, each individual tucked away in his or her own secret place.

Daytime is playtime for Oki and her new friends. A windy day serves as a climbing wall and a super-slide rolled into one. Up, up the young birds fly, struggling against the onrushing force of air. Then, with a quick flip, they catch the wind under their wings and allow themselves to be swept across country, tumbling after one another

and shouting at the top of their lungs. When the wind drops them back down near the ground, they leap into the air and climb back up for another ride. Again! Again!

One day, Oki notices a whippy branch dancing in a strong breeze and grabs hold of it with her beak. Soon, a small group of her playmates joins in, and they all hang there together, their bodies dangling and swaying like laundry in the wind.

Another time, she sweeps down to land in a blossoming tree with such a flourish and a bounce that a petal comes loose and falls. It lands on her brother, who is perched on a lower branch, making him jump with surprise. Hmm, interesting. Oki hops toward a blossom, plucks a petal with her beak, sidles back into position, and drops it on her brother's head. He jumps again, gives her a look, and moves out of reach.

"[Non-human animals] manifestly feel pleasure and pain, happiness and misery. Happiness is never better exhibited than by young animals… when playing together, like our own children."
CHARLES DARWIN, 1871

For scientists, play is a serious business. They interpret it as an important preparation and practice for adult life, a way for young animals to build muscle, sharpen their reactions, grow new connections in their brains, and generally learn about the way the world works. But how does an animal *feel* when she or he plays? When Oki drops the petal on her brother, does she think it's funny? When she and her friends come screaming down the wind, are they having fun? Some scientists shy away from asking these questions. They argue that we cannot know the answers and worry that we will be tempted to fill the empty space of Not Knowing with easy assumptions. *This is what I would be feeling, so crows must be feeling that too.*

But other researchers point out that refusing to consider these questions sets people apart from other species and doesn't recognize that we are all buds and branches on the great tree of life. For crows, the space of our Not Knowing is filled with their inner life. Just because we can't share it doesn't mean it isn't real.

CROW LAB

Keeping Score

Here are a few games that young crows sometimes play. See how many of them you can observe. The best time to watch for these behaviors is midsummer to early fall, when the juvenile crows are out exploring. Some of these games are common, others rare, so it could take a lifetime of crow-watching to see them all.

PRO TIP
Keep your records in your field notebook so you can add to them in the future.

▷ **Drop-and-Catch**
In this game, a crow picks up an object (a piece of bark or a pine cone, say), drops it, and swoops down to catch it.

▷ **Tug-of-War**
One young crow picks up a stick or a bone and another runs over to grab it.

▷ **Race-and-Chase**
A crow hops toward a companion, who runs or flies away. The chase is on—and quickly forgotten.

▷ **Bombs Away**
A young crow drops an object on someone else and watches to see what happens.

▷ **Wanna Fight?**
A young crow rolls onto its back and waggles its feet, inviting one of its playmates to wrestle. Adult crows sometimes have real fights over territories, but tussles among youngsters are always brief and playful.

▷ **Aerial Super-Sliding**
In this sport, young crows abandon themselves to the wind, tumbling and shouting as they stream cross-country.

▷ **Snow Sports**
Crows lie on their backs or bellies and slide down snowy slopes and rooftops or plunge their heads into drifts of loose snow.

Syrinx

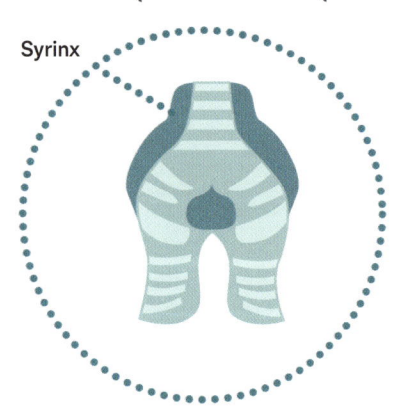

Syrinx

Certain songbirds, including some of the crows' closest cousins, can make two sounds at once, one from each side of the syrinx.

Being part of a swooping flock can be exhilarating, but Oki also spends time alone. Sometimes, she tucks herself into a quiet nook, low in a tree, and plays around with her own voice. She even sings to herself as she goes through her daily routine, probing in the ground or flipping through fallen leaves. As a fledgling fresh out of the nest, she was able to growl and gargle and cry for food, but that was about all. Then one day, she opened her beak and out came a rough caw.

Now she is able to ramble through a surprising range and variety of sounds: soft coos, rat-a-tat-tat rattles, clicks, moans, and a lot of subtly different caws. The caws may be long and drawn out, or short and staccato, or somewhere in between. Some buzz with vibrato, the way an opera singer would caw. Others are clean and clipped, like stone pinging against metal. Oki strings these sounds together in combination, either repeating one call over and over before switching to the next or streaming through them in a free-form improvisation. This murmured river of sound can go on for more than half an hour.

Like a young child who is learning to talk, Oki is gaining the ability to control her voice. Except for birds, most vertebrate animals—from lizards to lemurs to us—have an organ called a larynx [**lair**-inks], or voice box, that we use to make sounds. Birds, by contrast, have come up with something special and unique. Instead of a larynx, they have a double-barreled structure called a syrinx [**sear**-inks], which sits in the throat and upper chest like an upside-down letter Y.

Learn the Local Lingo

Spectrograms are computer-generated graphs that show the "shape" of sounds. By comparing the spectrograms of calls from different crows, scientists have discovered that individual crows have distinctive voices. This is likely one of the clues that crows use to recognize their family members and friends.

Can you tell the crows in your neighborhood apart by voice? If you can, you are a "rara avis" (rare bird)! The best most of us can hope to do is to distinguish a female, with her higher-pitched voice, from her lower-pitched male companion.

Something we can all do is to listen and watch. Notice the kinds of sounds your local crows make. What are they doing when they call? Are they alone or with others? Does anyone respond? What messages do they seem to be conveying? Do they always make the same sounds in the same situations, or do their vocalizations vary? Keep records of your observations. Maybe you will be the first person to break the code and fully understand crow-talk.

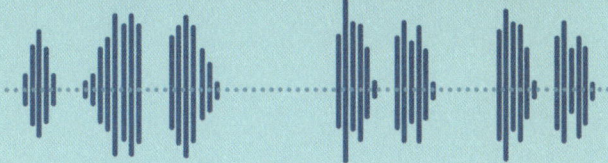

Caw caw caw ... ga-ga, ga-ga

Every time she caws, Oki draws air into her lungs and puffs it out through the twin pipes of her syrinx. Muscles flex; flaps vibrate; membranes rise and fall. Bony rings within her throat slide up and down over one another. The mechanism is so complicated that, even after years of study, scientists are still not entirely certain how it works. We do know, however, that every adjustment of the apparatus affects the sound that we hear, whether the caw is high or low, loud or soft, raspy or clear. These kinds of subtle differences allow crows to "say" a lot of different things. For instance, they can

- keep in touch with one another. *It's me. I'm over here.*
- issue commands. *Come here! Get away! Feed me!*
- defend their territories. *This is ours, not yours!*
- warn one another of danger. *There's a house cat. A flying hawk! A Great Horned Owl!!*

Of course, Oki doesn't know or think about the skills she is building as she warbles to herself. She just warbles away and carries on with her life. But through her mutters and murmurs, she is fine-tuning her ability to express herself clearly to her family and her growing circle of friends. That circle is about to get much larger as her world expands. Brace yourself for a great big noisy crow caw-palooza and confab.

CROW LAB

The Crow Said, "~!@#$%&*!"

Write an adventure story in which crows and people speak to one another in English. Then replace what the birds have to say with strings of nonsense symbols and letters. How does the story change? Can your crow and human characters still understand one another? Do they use their bodies or the tone of their voices to communicate?

Cawing 101

As a rule, the calls of songbirds are easy for us to understand. We know that a chirp means "here I am." A squawk signals alarm. A burst of song tells us that a male is defending his breeding territory. By contrast, tuning our ears to the calls of crows is wonderfully difficult. Their voices may not be tuneful, but they are remarkably flexible and expressive. When a crow says *caw*, its note may be loud or soft, long or short, low or high, flat or full-voiced. These subtly varied caws are often strung together, two by two or in long, gusty outbursts. The syllables may be timed to a steady beat or uttered in a great hullabaloo of sound. What does it all mean? Scientists still aren't entirely sure, but everyone agrees that it is well worth figuring out. Listen up!

Structured caws
Often, a crow will sit by itself on a conspicuous perch and utter a series of precisely timed caws. He or she then stops, waits, and repeats the series, usually reproducing it exactly. Structured caws are thought to signal that a crow is on his or her territory. They may also function as a call for the family to come together to fight off intruders or share in a newly discovered feast.

Alarm calls
When a crow notices a predator, she or he shouts in alarm. The greater the threat, the louder, faster, and more intense the warning becomes.

Squalling calls
Mayday, Mayday! A crow in the clutches of a predator utters a series of shrill, desperate-sounding screams. Other crows flock to this call and do what they can to intervene.

Rattles
If you hear a woodpecker tapping and tapping, like it doesn't know when to stop, you may be listening to one of the crow's rattle calls. Some rattles are soft and may serve to bring family members together or to signal playfulness. Others are harsh—like a stick clattering on the pickets of a wooden fence—and mean "keep your distance."

Coos
These quiet moaning calls sound a bit like a pigeon cooing. They seem to be associated with peaceful interactions in the family.

Body Language
Crows add extra drama to their communication through body language. For example, birds who are feeling confident and assertive stand tall, tilt their bill upward, and ruffle the feathers on their throat and legs. These birds strut around, cawing loudly. Birds who are submissive lower their gaze, sleek their plumage, and slink away.

Crows can add intensity to their vocalizations by bowing deeply with each caw. Or they flick their wings and fan their tails, adding a bit of razzle-dazzle that shows they are agitated or excited.

6

Life Choices

Summer only
Year round

American Crow range map

The circle of the seasons continues to turn, and the days are again getting short. An icy wind has blown in from the north, sending gusts of dried leaves skittering across the grass. Skeins of geese silver the sky, and flocks of crows rise and swirl in the moving air. Some of those restless fliers—including most of the crows that spend their summers in Canada—will keep shifting south for days or even weeks before finally choosing a place to settle and wait for spring. Other crows—including those that live on the coasts and at midlatitudes in the United States—are able to stay closer to home. But even for them, winter is a season of change.

The luxuriant green world that sustained Oki and her family through spring and summer has turned brittle and bare. There are no frogs, no juicy worms or grubs, no fresh berries. Mice are hidden away in their burrows, and the buzz of flying insects has been stilled by the cold. The leafless trees no longer provide protection from the sharp eyes and sharper talons of Great Horned Owls.

One afternoon, Oki and her family are foraging in a stubble field with some of their familiar companions, busily picking up weed seeds and kernels of spilled grain. The sun is sliding down the western sky and the shadows are growing long. With a whir and rustle of wings, individuals and small groups of crows leap into the air and fly away in one direction or another. Ordinarily, this would be a cue for Oki to head to the summer roost-trees or back to her home territory with the rest of her family. This time, however, she is stirred by a strange new desire. When one of her familiar caregivers heads off in an unfamiliar direction with a gang of unfamiliar crows, she leaps into the air and races to join them.

The flock sweeps across town and country and then swoops down—a cloud of black shapes descending—to land on a wide expanse of lawn. Hundreds of crows are already gathered there, feeding and playing on the ground or chasing one another in circling flights around the treetops. As the sun slips toward the horizon, groups of crows begin to tear off from this large gathering and stream away, one ragged throng after another, all zooming in the same direction.

By now, Oki has lost sight of her companion, but if she is worried, it doesn't show. She is carried along in the excitement as more and more crows sweep toward the same destination, an open area with plenty of places to perch. This is the crows' communal winter roost. By day, the area serves as a big-box shopping mall. By night, it is a city of crows.

When darkness finally falls, the flock of hundreds has grown to thousands and tens of thousands—screaming, squabbling, pooping—all gathered together within a few city blocks. (Next morning, a crew of workers will come by with a power hose to clean up the mess!) With streetlights to see by and thousands of eyes to keep watch, the crows settle in shoulder to shoulder to get what rest they can in the constant fuss and flutter of this all-night confab.

Give a Roost a Boost

A winter crow roost is a wonder of nature, but it can also be a giant, noisy nuisance. Just ask the family of crows that occupied a peaceful territory on the campus of Cornell University in Ithaca, New York. Everything was fine until, without warning, waves of outsiders, as many as 2,000 in all, started dropping in, uninvited, for a slumber party. The resident crows screamed at the intruders and dived and chased, but they didn't have much effect. Eventually, the roost grew too large for the territory, and the party shifted to a new place.

Much the same thing happens when a crow roost forms in a city or town. Disgruntled human residents try everything they can think of to persuade the birds to leave, from blasting them with loud noises to blasting them with guns. But even lethal force is generally not very effective in breaking up large roosts.

Instead of fighting the crows, some communities, like Burnaby, British Columbia, and Danville, Illinois, have chosen to celebrate their presence as a way to attract tourists. Standing in the midst of thousands of wheeling, calling crows is an unforgettable experience. Is there a large crow roost somewhere close to you? Could you persuade your family to go for a visit?

PRO TIP
If you don't know where the closest winter roost is, try contacting your local birding or nature society for information.

A crow can fly up to 50 miles (80 km) in a single day (25 miles or 40 km there and back), scanning the landscape for anything edible.

When the new day dawns, Oki will be faced with choices she has never made before. She could decide to fly back home and spend the day with her parents on their territory. Or she might choose to take full advantage of her sudden independence and tag along after one of her new acquaintances. A crow who leaves the roost early, on a direct, unswerving course, probably knows where to find food. Why not take a chance and follow along? The reward might be a bird feeder stocked with hard-boiled eggs or a dead skunk in a ditch. It might be a dumpster overflowing with scraps of pizza or a crop of pecans spread under a tree.

Back when she was a nestling, Oki's diet consisted mostly of insects and other animal-based foods, but with age, her tastes have become more omnivorous. Her menu now includes more grains, nuts, and fruits. Every resource the land has to offer is entered on Oki's mental map, as the circumference of her world expands.

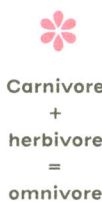

Carnivore
+
herbivore
=
omnivore

CROW LAB

Mapping the Known World

Working from memory, make a map of your neighborhood, showing all the places that matter in your life. Where do you relax and sleep, learn and play? Where do you find shade or pleasant views? Where do you get your food? Be sure to include your garden, if you have one, and your favorite ice-cream stands and restaurants, as well as the grocery store. How large an area does your map cover?

Now try this project again as if you were a crow. Map the location of large trees that could provide shelter for nesting. Where are the open lawns and fields where crows can forage? Be sure to include gardens, bird feeders, fast-food malls, and garbage dumps. How large an area does the crows' map represent?

How do the two maps compare? Are there places that are important to both you and your crow neighbors?

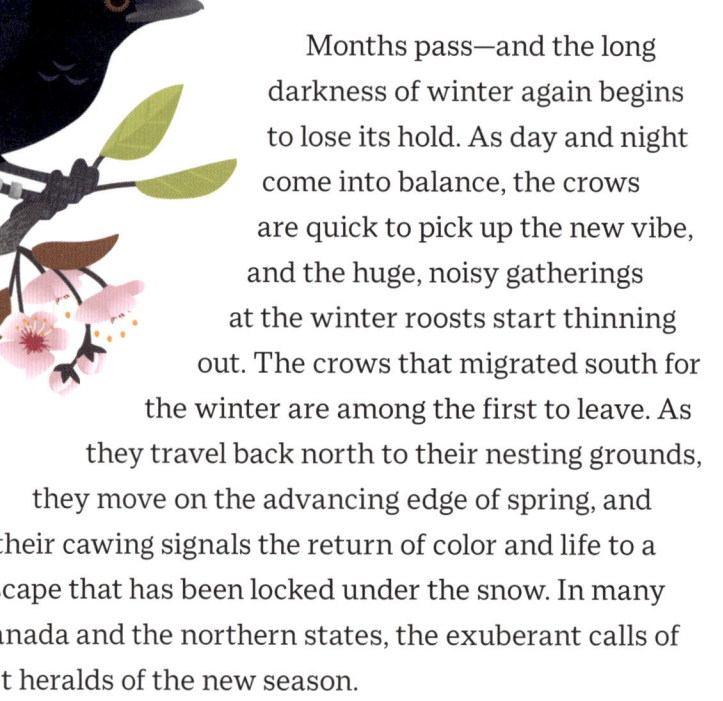

Months pass—and the long darkness of winter again begins to lose its hold. As day and night come into balance, the crows are quick to pick up the new vibe, and the huge, noisy gatherings at the winter roosts start thinning out. The crows that migrated south for the winter are among the first to leave. As they travel back north to their nesting grounds, they move on the advancing edge of spring, and their cawing signals the return of color and life to a landscape that has been locked under the snow. In many parts of Canada and the northern states, the exuberant calls of crows are the first heralds of the new season.

For Oki, the lengthening days of late winter are a signal to spend more time back on her home territory. Soon the whole family has reassembled. Her three broodmates are there, as are the helpers who cared for them when they were nestlings. Meanwhile, her parents are already prospecting for a place to build their new season's nest. Sometimes, they perch in a tree and gaze up at the intersecting branches, as if assessing the possibilities. They have started breaking off twigs and carrying them around in their beaks. All the signs seem to be pointing toward another successful year.

Robins are often celebrated as the first birds of spring. But in many places, crows are the real early birds.

But then something unexpected happens. Oki's mother disappears. She leaves to feed at the garbage dump, as she has many times before, but this time she doesn't return. (Don't worry! She'll be all right. More about that later.) And meanwhile, life carries on. Almost immediately, a new adult female arrives and starts to cozy up to Oki's father. For the sake of convenience, we'll call her Princess 1. She had begun her life on the other side of town but has been hanging out with a family nearby for the last couple of years. Now here she is, sitting side by side with Oki's father, as they preen each other's feathers and hold one another's bills, lovey-dovey as they come.

But it is not to be. Before the new pair can begin nesting, a second adult female, Princess 2, turns up and chases Princess 1 away. Older and more experienced, the newcomer has been a helper for five nesting seasons and is more than ready to start raising her own young. So now it's Princess 2 and Oki's father who are preening each other's feathers and holding one another's bills. Together, they will produce a brood of three nestlings.

But Oki will not be around to meet the hatchlings, because early on in this kerfuffle—shortly after her mother's disappearance—she slips away from home. One of her broodmates, a sister, makes the same decision. Independent of each other, they spend the spring and summer hanging out with other families or loafing around with other unattached yearling crows.

Oki is now one year old. *Cawngratulations,* Oki. Way to grow!

Game of Crows

BY AND LARGE, CROWS GET ALONG REMARKABLY WELL WITH ONE ANOTHER. BUT UNDER THE PEACEFUL SURFACE, HIDDEN TENSIONS LURK. IMAGINE...

FOR MANY YEARS, THE OLD KING AND QUEEN HAVE REIGNED HAPPILY OVER THEIR TERRITORY, SURROUNDED BY FAMILY AND FRIENDS. THEN, OUT OF THE BLUE, DISASTER STRIKES.

THE QUEEN IS KILLED IN A GREAT HORNED OWL ATTACK!

THE FAMILY GATHERS ROUND...

BUT WHO'S THIS? A PRINCESS ARRIVES FROM AFAR AND FINDS FAVOR WITH THE KING.

A RIVAL PRINCESS FLIES INTO THE FRAME AND CHASES THE OTHER PRINCESS AWAY.

THE KING AND HIS NEW QUEEN ARE CONTENTED

BUT TWO OF THE OLD QUEEN'S DAUGHTERS DECIDE TO TAKE THEIR LEAVE.

MEANWHILE, ONE OF THE KING'S SONS WAITS TILL HIS FATHER IS NOT LOOKING AND SNEAKS CLOSE TO THE NEW QUEEN.

WHAT IS HE UP TO? THE PLOT THICKENS... (WINTER IS COMING.)

One of Oki's favorite hangouts is the local compost dump with its bounteous supply of deliciously rotting stuff. Usually, she finds herself surrounded by random acquaintances, but one day, she looks up and spots her sister. They rush to meet each other and spend a few minutes preening each other's feathers. Then they fly off, each in her own direction, and go back to feeding.

Oki is still a youngster, but she has already led a life filled with intimacy and independence, excitement and risk. What might lie in store for her as she crosses the threshold into adulthood?

These stories are not made up. They are based on careful observations by scientists who study marked birds. It's a fact! Crows are amazing!

A Season of Crows

Haiku is a Japanese form of poetry that has been perfected over hundreds of years. In writing a haiku, the goal is to convey a strong image and emotion with just a few words. A classical haiku usually refers to the seasons and consists of three lines. The first and third lines are limited to five syllables, while the second expands to seven syllables or sometimes to nine. Try for a break in the thought at the end of the second line to heighten the impact. All these rules and restrictions add to the fun and satisfaction of capturing your experience in words. Here is an example to get you started:

Blue shadows deepen
Spent leaves rustle in the wind—
Listen to the crows

A Season of Dying

To see Oki now, it is hard to believe that she was once a bedraggled hatchling struggling to get free of her egg. As she nears her second "hatch-day," she is elegant and alert, with shining brown-black eyes, a shining black bill, and shining black feet. In the low-slung sunlight of late winter, her plumage shimmers with glints of purple and green.

Apart from the disappearance of her mother and, before that, the loss of a sibling in the nest, her progress through life has been smooth. She has moved through all the ages and stages of youth—from hatchling to nestling to fledgling to free-flying explorer—without a hitch. Now she is poised to make the transition into adulthood.

American Crows become sexually mature at the age of two. In other words, two-year-olds like Oki are physically capable of mating, producing fertile eggs, and raising young. As a rule, though, they do not rush into parenthood. Females usually breed for the first time when they are three or four years old, and sometimes later than that.

By the age of two, a crow has dodged many risks. He or she has a good chance of living another eight to ten years— or even longer.

The oldest wild American Crow on record lived to be 29 and a half years old.

Males wait even longer and typically father their first broods at the age of five or six. Even then, these young-adult breeders are inexperienced, and their early attempts often end in failure, with tumbledown nests, infertile eggs, or drooping, malnourished chicks.

One alternative to breeding on your own is to stay home and help your parents raise their next brood. That's what a lot of crows do when they are yearlings (between their first and second birthdays). It's what Oki and her sister might have done if their life hadn't been disrupted when their mother vanished. Their brothers did choose to remain on their home territory through all the confusion and change. They even pitched in to help their father and Princess 2 raise the year's nestlings.

This stay-at-home behavior is common among young-adult males. When a male is finally ready to breed on his own, he typically takes over part of his father's territory or settles down nearby. A female,

by contrast, is likely to go on an adventure, possibly ending up many miles from where she hatched. But no matter where she goes, she will never forget where she started out in life. Sometimes, after an absence of months or even years, she may decide to return for a visit.

Perhaps that is what Oki is doing on a fateful day in late winter when, at the age of almost two, she takes a notion to fly to her home territory. She has been away for many months, ever since the kerfuffle with the Princesses. Will her family welcome her back and let her help with the upcoming breeding season? Teetering on the top of a tall spruce tree, she shouts out a loud hello and then listens for a response. But no matter how long and lustily she calls, no one answers.

Where has everyone gone? Her father, Princess 2, her brothers, the helpers who brought her up, the young of the year—her entire family is gone. What could possibly have happened?

True story:
An adult, nesting female was badly injured by a hawk. She flew home to stay with her parents for several months.

The next spring, she returned to her breeding territory, reunited with her mate, and went back to nesting.

To their sorrow, the scientists who have been keeping an eye on Oki's family know exactly what is going on. They have watched the birds die, one after another. From testing their remains, the researchers can tell that the crows were struck down by a runaway virus. West Nile virus has

Attend a Crow "Funeral"

When a member of a crow family is dying, the others in the group usually sit nearby and keep a silent watch. Occasionally, one of them may bring food or sit shoulder to shoulder with the sufferer, offering support. But if a crow discovers the dead body of a stranger, that is an entirely different matter. *Caw-caw-caw*—wild outbursts of agitated shouting bring crows streaming in from far and wide. From the commotion, you might think the crows were confronting a terrible danger, like a Great Horned Owl. But no, when you get there, you can see they are shouting at a clump of black feathers, lying still on the ground. Some researchers speculate that the crows are trying to figure out what happened, and that they have a concept of death. Everyone agrees that these gatherings are remarkable and mysterious.

existed in parts of Africa and the Middle East for many years, without causing serious concern. But in 1999, it showed up way across the ocean in New York City (no one knows quite how it arrived), and scientists started to get worried. Within three years, the virus had spread across North America—north, south, east, and west. Soon it was popping up almost everywhere.

West Nile virus primarily infects birds. It is transferred from one bird to the next and the next in the saliva of mosquitoes. Some species of mosquitoes are specialists that bite only birds, but others seek out a wider range of victims, including both birds and the enticingly naked flesh of humans.

If a mosquito that carries West Nile virus sticks its proboscis into a person, there is a chance that person will become seriously ill, though fortunately this is rare. In humans, serious illness occurs in fewer than one case in a hundred. A larger group of people (about 20 in a hundred) will experience fevers, headaches, and other mild symptoms. But most people who get infected with West Nile virus (80 out of that same hundred) don't feel sick at all.

Health for One, Health for All

West Nile virus is zoonotic [zoo-oh-**not**-ik], meaning that it affects wildlife and people. By helping to prevent the virus from spreading, we can protect ourselves and other creatures, including our crow friends. Here are a few important things to remember.

- If people are infected with West Nile virus, they can get seriously ill, but fortunately this is rare.

- West Nile virus is only spread to people by a few kinds of mosquitoes that bite both birds and mammals.

- You can protect yourself from mosquito bites by wearing light-colored clothes that cover your arms and legs. You can also use insect repellent on the parts of you that poke out, but be sure to wash it off when you go back indoors. Repellent is especially helpful if you are outside at dusk, when these mosquito species are most active.

- Mosquito larvae, or "wrigglers," develop in still, shallow water. If you are lucky enough to have a birdbath, take care to change the water every few days. Don't leave water standing in plant saucers, forgotten buckets, or paddling pools.

- If you are super-duper-lucky enough to have a horse, make sure your horse is vaccinated. (Yes, there is a vaccine against West Nile virus for horses, but not for people or birds!)

- Because crows are so sensitive to the West Nile virus, scientists use them as a "sentinel species." When lots of crows die, it is a sign that there is lots of the virus around. If you find a dead crow, please do not pick it up. Ask an adult to help you report your discovery to your local wildlife office or public health authority. Sad as it is to find a dead crow, it is satisfying to help scientists understand and track the virus.

Happily, many species of birds come through a West Nile infection without suffering great harm. But this is not true of corvids. It is not true of American Crows. If an infected mosquito bites a crow, the virus usually takes hold fast. Within a day or two, the bird stops eating and drinking. It sits and stares. It may twitch and flop around on the ground, unable to control its body. Other members of its family perch quietly nearby, keeping watch, but there is nothing they can do. Within a week of that fatal mosquito bite, the sick bird will be dead. The fatality rate for American Crows infected with West Nile virus is close to 100 percent.

Crows have no resistance or natural immunity to the West Nile virus, and there is no vaccine to protect them. In bad years, when warm, moist weather produces clouds of mosquitoes, crows may die by the hundreds and the thousands. The worst time for them is late summer, and that was when the members of Oki's family were struck. But, of course, she doesn't know this. All she knows is that they are not waiting to greet her where she thought they would be. So strange. So confusing.

But Oki has her own life to live—and spring is calling!

CROW LAB

Something to Crow About

Crows who are infected with West Nile virus sometimes recover if they get enough TLC (tender loving care). Caring for sick and injured animals is the job of veterinarians and licensed wildlife rehabilitators. Is there a wildlife rehab clinic near you? Could you make a donation or raise funds for them? Imagine people flocking to purchase your handmade crow cards or cookies or crafts. Imagine them lined up to buy tickets for a crow-themed talent show or cawing competition or dance performance. Get creative. Helping birds and animals get the care they need is something to crow about!

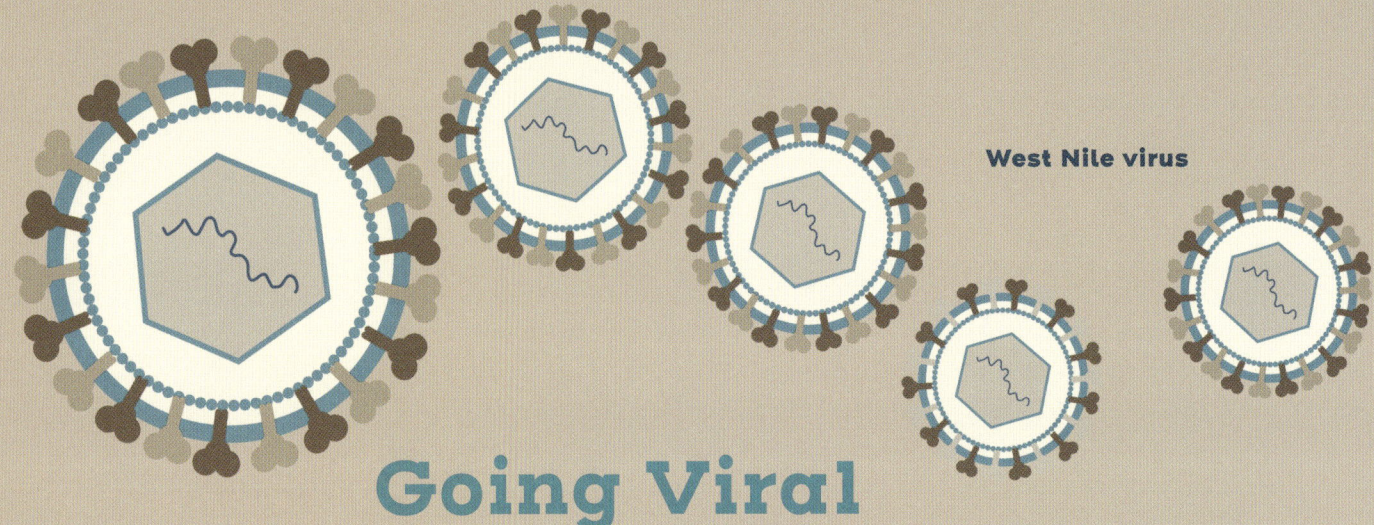

West Nile virus

Going Viral

West Nile virus circulates mainly between birds and mosquitoes. Birds amplify the virus, meaning that they can build up high levels of virus in their bloodstreams. Female mosquitoes need a meal of blood in order to lay their eggs. When a blood-hungry mosquito bites an infected bird, she picks up the virus. From then on, she will pass it on to any creature she bites, whether it is another bird, horse, or human. Keeping crows and other wildlife healthy helps to keep people healthy too.

Unlike the virus that causes COVID-19, West Nile is not transmitted through the air. What's more, it does not pass easily from one person (or horse) to another, because we do not amplify the virus. Since we don't transmit the virus to others, we are dead-end hosts for West Nile. Although the virus makes some people very sick, it does not cause runaway outbreaks among humans.

**Amplification host
(Birds)**

**Incidental host
(Humans)**

WEST NILE VIRUS
Primary Transmission Cycle

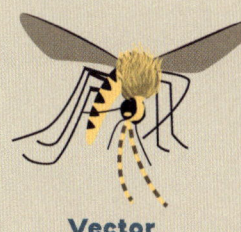

**Incidental host
(Horses, etc.)**

Vector (Mosquito)

A Nest of Her Own

Sometimes newcomers to a group are welcomed, and sometimes they are not. It's all very personal.

The new season floods over the land like a tide of joy. Everywhere, fruit trees are bursting into blossom, and wave after wave of returning birds are raising sweet voices in song. And it is at this time of wonder that Oki makes an amazing discovery.

It is just over a year since her mother disappeared, and the unfortunate (now dead and departed) Princess 2 moved in to fill the gap. As a yearling, Oki had taken this upheaval as a cue to leave her home territory and go exploring. Since then, she has spent much of her time with a family of crows across town. Not that they have exactly welcomed her in with open hearts. The breeding female, in particular, is touchy and aggressive. Anytime Oki dares to come close to the nest site—whether with a beak full of nesting materials or a crop full of food—this mother-crow-who-is-not-Oki's-mother shouts at her and sends her packing.

All the same, the family doesn't seem to mind if she hangs around the edges of the territory and forages along with them. And she is also allowed to help out as a sentinel, or lookout, and to join in the frenzied commotion if—terror of terrors!—a Great Horned Owl appears.

The more we learn about the intimate lives of crows, the more astonishing they become.

So it happens that, one fine spring morning when Oki is two years old, she and some of her new acquaintances fly out to a freshly plowed field together, to feast on earthworms and grubs. Dozens of crows have already gathered there, some swooping and calling in the air, others on the ground, heads down, beaks probing, intent on breakfast. That's when Oki hears a familiar voice and sees a familiar form. One of the crows feeding in the field is *her mother*.

Of course, Oki will never know the whole story. But it turns out that, on the fateful day of her disappearance, Oki's mother had been injured by a kid with a pellet gun. (Taking potshots at crows is against the law in many places, but it still sometimes happens.) Fortunately, a thoughtful person was able to catch her and take her to an animal clinic for care. And so here she is—Oki's mother—bright and shiny as ever. Reunited, mother and daughter sit side by side and preen each other's heads. Then they zoom off together to the mother's territory, just a few minutes away. This time, Oki's offers to help at the nest are welcomed. She is allowed to feed nestlings, carry away their fecal sacs, and generally pitch in.

Share the Love

The Migratory Birds Convention is an agreement between the United States and Canada that protects migratory birds in North America from being deliberately harmed by people. Specifically, the agreement makes it illegal to pursue, hunt, shoot, wound, kill, trap, capture, or collect migratory birds—or even to try to cause these harms. Although American Crows appear on the list of migratory species, they are not fully protected. A special clause permits states and provinces to allow crows to be killed as pests (with an official permit) and to be hunted for fun (during a restricted hunting season). The only crow species that is completely protected is the Hawaiian crow, *Corvus hawaiiensis*, which is endangered.

People who shoot crows for sport think of the birds as disease-ridden varmints. They claim, falsely, that crows are carriers of West Nile virus. In fact, infected crows are so susceptible to West Nile virus that they quickly *die* of the disease. They are unlikely to live long enough to amplify it and pass it on to others.

It is important to spread the word that crows are smart, sentient creatures who live in families and lead complex social lives. They should not be viewed as pests or used for target practice. Find out how crows are treated where you live. Share what you learn. Make a poster. Write a poem. Tell a story. Create a display for your science or heritage fair. Shine a light on the lives of the crows that live around you.

Busy as she now is, Oki still makes time to stay in touch with her adopted family. She regularly drops by their area to forage and help keep watch for intruders. But a new desire is stirring in her body, an unfamiliar energy that is changing the way she behaves. More specifically, this agitation is making her exceedingly noisy. Sometimes, she sits on a high perch on the edge of her friends' territory and caws.

And caws.

And caws.

Here I am, she hollers. *I'm all grown up now.*

Listen to the Music

The cawing of crows may not be tuneful, but it has a music all its own. Listen: Can you hear the beat? Can you feel the emotion behind it? Does it remind you of beatboxing or Inuit throat singing? How can you capture the rhythms and intensity of crow song to make your own music?

When she isn't shouting, Oki likes to hang out with a particular member of her adopted group, a good-looking three-year-old male who is living on the territory with his family. The two young adults often loaf around together, and early in the season, they spend a couple of weeks working—in a half-hearted, on-and-off kind of way—at building a nest, only to lose interest and abandon it.

By late summer, they are often seen feeding and perching in an area on the edge of the territory, apparently trying to claim it for themselves. This process is called "budding" and it often works out well, allowing new pairs to settle close to their parents. But things aren't going smoothly for Oki and her friend. The next-door neighbors are aggressive and keep driving the young pair off. To make matters worse, the young male's mother (the mother-who-is not-Oki's-mother) is as grouchy and intolerant as ever. Laying claim to a territory will have to wait for another season.

Living well with friends and foes takes a lot of figuring out. Animals that are very sociable are usually also very smart.

And so, summer gives way to another winter. Oki spends the short days at home with her mother's family and the long nights in the sociable excitement of a communal roost. But one day, as the light is seeping back into the world and the trees again are greening with the promise of spring, she takes a notion to fly over and visit her adopted family. If she is expecting to find her "boyfriend" from the previous summer, she will be disappointed. He has gone off to live with an older brother who is breeding a short distance away. Why? Only he knows for sure. But the short answer is that he left because he wanted to. He saw an opportunity and decided to act on it—which is exactly the kind of life-changing choice that Oki is about to make.

The young male is not the only individual who is absent from the group. His mother (the crotchety breeding female-who-is-not-Oki's-mother) has also disappeared, succumbing to old age in the hard months of winter. Could it be that this is what had attracted Oki to this family all along? Had she noticed that the breeding female was no longer young and deduced that a vacancy might soon open up?

Oki is ready and willing to take on the challenge. The next thing you know, she and the widowed male are sitting together, wooing one another with soft coos, rattles, and clicks. They bow and groan and fan their shining tails. The male is seven years old and experienced. Oki is now fully grown up and resplendent at the age of three. Whatever else adult life may hold in store, this is a good beginning.

"The course of crow love never did run smooth."
WILLIAM SHAKESCROW

Extraordinary Lives

American Crows often lead complex lives, full of drama and intimacy. Their stories are shaped both by what is happening around them and by their own personalities and choices. Here are short biographies of one family group of marked crows. As they experience life's ups and downs, do they feel love and grief, frustration and happiness? Imagine...

QR—Breeding male. Quincy budded a territory from his parents, then formed a bond with a female, TX (Talullah). When she died, he paired with Olive. Eventually, he will be displaced by his son Xander and will settle with a widowed female nearby.

OM—Breeding female we'll call Olive. Olive lived with her parents until she was three, then paired up with QR (Quincy). Later, she will partner with XL (Xander) and produce many more broods of young.

XL—Adult son of Quincy and Talullah (Quincy's previous mate). Xander stays home with his father and Olive. At age six, he will displace Quincy and take over as breeding male.

XG—Daughter of Quincy and Olive. Xena will leave home during her first winter and eventually settle down a long way from home. She will return occasionally for brief visits.

PZ—Son of Quincy and Olive. Pablo will stay home until his father leaves, then move to Quincy's new territory. At age six, he will find a mate, bud off a territory of his own, and raise his first of many broods.

KI—Daughter of Quincy and Olive. Kyrie will stay home for two years, then choose a mate nearby. When her mate dies of West Nile virus, she will pair with a widowed neighbor, adopt his brood of young, and care for them as her own.

And they will all live as happily as possible under the circumstances!

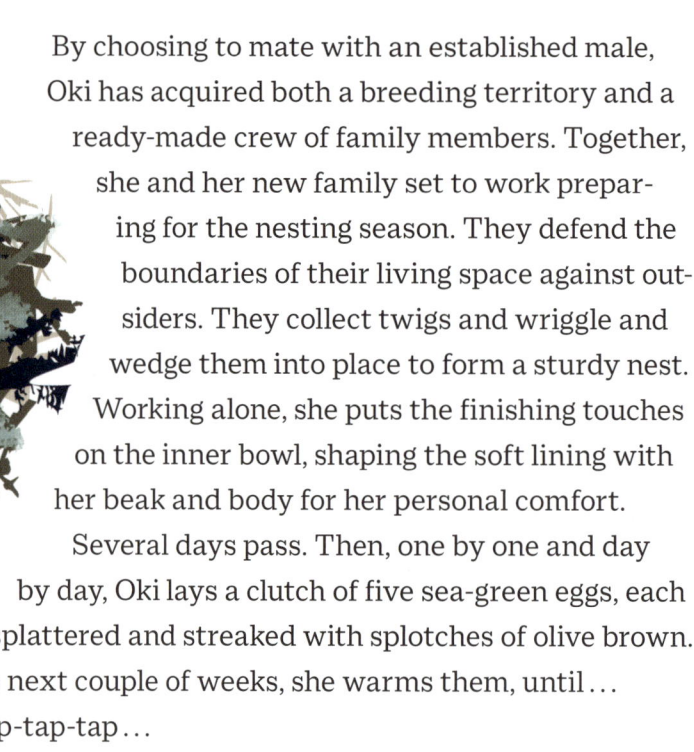

By choosing to mate with an established male, Oki has acquired both a breeding territory and a ready-made crew of family members. Together, she and her new family set to work preparing for the nesting season. They defend the boundaries of their living space against outsiders. They collect twigs and wriggle and wedge them into place to form a sturdy nest. Working alone, she puts the finishing touches on the inner bowl, shaping the soft lining with her beak and body for her personal comfort.

Several days pass. Then, one by one and day by day, Oki lays a clutch of five sea-green eggs, each splattered and streaked with splotches of olive brown. For the next couple of weeks, she warms them, until…

tap-tap-tap…

something, or someone…

deep in the nest…

starts stirring.

Like her own mother three years earlier, Oki rouses from her slumber and perches on the rim of the nest. The tip of a pale beak is barely visible through a fissure in the shell. Within days, one, two, three, four, five, the nest will fill with a tangle of scrawny limbs and grape-round abdomens. The hatchlings will be naked, blind, helpless, and so very hungry.

From her vantage point on the rim of her nest, Oki looks down at the tiny creature emerging at her feet. Then she looks up and out at the view, measuring the circumference of her experience. The horizon spirals around her, zooming over streets and rooftops to embrace the woods and fields (and dumpsters) that she has come to know so well. Within her gaze lies everything that she and her offspring will need to survive and thrive: food, shelter, acquaintances, family, and friends. And people like you and me who care about her.

So long for now, dear Oki. See you Out There!

Find Your Flock

Caring about wildlife in a world of troubles can feel lonely and stressful at times. But remember: you are not alone. If there is an ecology club in your school, join it. If one doesn't exist yet, ask a friend to help you get something started. Reach out! Connect with your local nature society or young naturalists' program. If other members of your family share your love of birds, invite them to go exploring with you. Be communicative, be companionable, be curious. Channel your inner crow!

Author's Note

If this book were a cake, the recipe would read: "Take four cups of fact and one cup of fantasy, blend well, and bake."

The information in this book and the events in Oki's life story are all based on scientific research. If the mention of science makes you think of white coats, microscopes, and test tubes, you are not wrong. Some of the knowledge that is shared in these pages was discovered in laboratories. Figuring out the structure of the avian brain and calculating the impact of West Nile virus are two examples.

But the scientists who have contributed most to telling Oki's story prefer to drop their lab coats, grab their backpacks, and head for the great outdoors. Ever since they were children, these women and men have been madly in love with the living world. That passion led them to attend university and train as biologists. Exactly what motivated them later to devote decades of their lives to studying hard-to-catch, wild-spirited American Crows is a question that only they can answer. But I am very thankful that they took on this difficult challenge.

I am especially grateful to the three scientists who shared their hard-won knowledge with me. They are:

* Behavioral ecologist Dr. Carolee Caffrey, who studied the social behavior of individually marked populations of American Crows in Oklahoma and California and who is also the co-author of an important scientific account of the species;

* Professional ornithologist Dr. Kevin J. McGowan of the Cornell Lab of

Ornithology, who is a mainstay of a multi-decade study of American Crows in New York State; and

* Behavioral ecologist Dr. Anne Barrett Clark of Binghamton University. Together with Dr. McGowan, Anne Clark and her students have tagged and banded more than eight generations of American Crows around Ithaca, New York, and documented the twists and turns of their complex life stories. Anne made time for repeated question-and-answer sessions with me and also shared maps, references, photographs, and other resources. This book would not have been possible without her generosity and kindness. Her friendship is the icing on the cake of this project.

As for the element of fantasy that I mentioned earlier, I hope you will not be disappointed to learn that Oki is imaginary. Her story is pieced together from the experiences of actual crows, as observed by my scientific advisers. Everything that happens in these pages happened to a crow in real life. So even though Oki herself is a fiction, her story is filled with truth.

I am grateful to the Canada Council for the Arts for a grant that supported the creation of this book. It is also a pleasure to acknowledge the contributions of my colleagues at Greystone Kids. Thanks to Rob Sanders for inviting me to undertake this project, Linda Pruessen for her sympathetic editing, Dawn Loewen for her sharp-eyed copy editing, Tracy Bordian for her proofreading, and Jessica Sullivan for working her usual magic with the design. Natural history illustrator Rachel Hudson brought Oki's story to life by balancing precise observation with a confident sense of design.

Finally, a special word of thanks to my nearest and dearest—Keith, Diana, Marilyn, Asha, and Laurel—and to our neighborhood crows for their gift of wonder.

CANDACE SAVAGE

 # Glossary

Addled Mixed up. In the case of an egg, rotten.

Altricial Requiring extended care by parents.

Amplify Increase. In the case of a virus, multiply.

Auxiliary Helper, supporter, backup.

Brachyrhynchos Short-beaked or short-nosed.

Brood As a noun, a family of young. As a verb, to sit over eggs or young birds to warm them.

Bud Take over a portion of an existing territory.

Carnivore An animal that eats meat.

Cerebral cortex Wrinkled structure at the front of a mammal's brain.

Clutch Group of eggs laid by a female bird in one nesting attempt.

Contour feathers The outer feathers on the body of a bird. Includes wing feathers needed for flight.

Convergent evolution The process through which species that are not closely related to each other become similar, shaped by similar environmental challenges.

Cooperative breeding A family-based system in which older offspring remain with their parents and often help to raise their parents' young.

Corvid A member of the crow family (*Corvidae*), which includes crows, ravens, magpies, jays, nutcrackers, and others.

Corvus The large genus, or group, within the corvids that consists of crows, ravens, and rooks.

Darwin, Charles English naturalist and thinker, a founder of modern biology, who lived from 1809 to 1882.

Dead-end host An animal that is infected by a virus but does not pass it on.

Down Soft, wispy feathers.

Egg tooth A bump on the bill of a newly hatched bird, used to crack open the eggshell.

Embryo A developing animal before it hatches or becomes a fetus.

Excrement Solid waste eliminated from the body; poop.

Fecal sac Tidy little package of poop from a nestling bird.

Fetus An unborn animal that has developed the basic shape it will have when born, usually referring to mammals.

Fledge Leave the nest permanently.

Fledgling A young bird that has permanently left the nest.

Forage Search for food.

Forebrain The front part of the brain, often responsible for complex intelligence.

Gene Part of a molecule of DNA that influences the appearance or behavior of a living thing. Genes pass information, in a kind of code, from generation to generation.

Genus A group of closely related species. For instance, crows and ravens are members of the same genus, *Corvus*. Humans are in the genus *Homo*.

Gular pouch A space in the neck of some birds, used for carrying food.

Gullet Throat.

Hatchling A bird that has just hatched.

Herbivore An animal that eats plants.

Incubate In the case of birds, keep eggs warm so they will hatch.

Infertile Unable to develop.

Instinct An inborn, reflex action.

Juvenile Young, immature, not yet adult.

Larynx The voice box of a mammal.

Migratory Moving up and down the continent with the change of seasons.

Mob In the case of crows, gang up on a predator.

Murder An old-fashioned term for a group of crows. The preferred term is a "flock."

Nestling A baby bird before it has left the nest.

Neuron A nerve cell.

Nidopallium caudolaterale The front part of the brain of a bird.

Omnivore An animal that eats food that comes from both plants and animals (meat, eggs, etc.).

Pinfeather A feather just emerging through the skin, often encased in a sheath.

Pip Crack open the eggshell while hatching.

Plumage All the feathers that clothe a bird.

Precocial Not requiring much parental care. Advanced for its age.

Predator An animal that hunts other animals.

Proboscis In insects, a sharp tube for sucking or piercing.

Roost As a verb, to perch. As a noun, a place where birds gather to perch together.

Sentient Aware, conscious.

Sentinel A lookout.

Sentinel species A species that gives people a heads-up about environmental risks.

Songbird A member of a large group of birds, including wrens, sparrows, thrushes, and crows, that have a specialized vocal organ and the ability to produce complex vocalizations.

Species A unique kind of organism. Members of the same species can produce young.

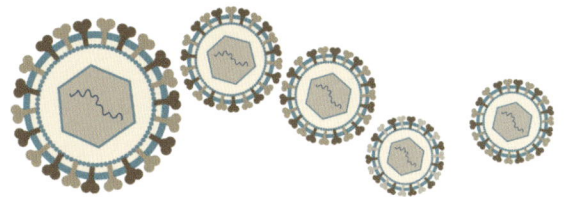

Spectrogram A special kind of diagram used to represent and study sounds.

Subspecies A subgroup within a species. Members of different subspecies could produce young together but rarely do so because they are separated by distance.

Synapse The connection between nerve cells.

Syrinx The complex voice box of birds.

Territory In biology, an area claimed by a pair or family for breeding and foraging.

Theropods Carnivorous dinosaurs that walked on two legs, including tyrannosaurs and velociraptors. The ancestors of modern birds.

Umbilical scar The mark on a newly hatched bird where it was attached to the egg yolk before emerging from the shell.

Vertebrates Animals with backbones, including amphibians, fish, reptiles, birds, and mammals.

West Nile virus A virus that infects birds, humans, horses, and some other animals.

Wildlife rehabilitation Specialized care for wild animals that are sick or injured.

Yearling An animal that is between one and two years old.

Zoonotic disease A disease that passes between animals and humans.

Syrinx

Resources

Books

Cate, Annette LeBlanc. *Look Up! Bird-Watching in Your Own Backyard.* Somerville, MA: Candlewick Press, 2013.

Oseid, Kelsey. *Nests, Eggs, Birds: An Illustrated Aviary.* California and New York: Ten Speed Press, 2020.

Pringle, Laurence. *Crows! Strange and Wonderful.* Honesdale, PA: Boyds Mills Press, 2010.

Savage, Candace. *Bird Brains: The Intelligence of Crows, Ravens, Magpies, and Jays.* Vancouver, San Francisco: Greystone Books, 2018.

———. *Crows: Encounters With the Wise Guys of the Avian World.* Vancouver, San Francisco: Greystone Books, 2015.

———. *Hello, Crow!* [picture book]. Vancouver, San Francisco: Greystone Books, 2019.

Turner, Pamela S. *Crow Smarts: Inside the Brain of the World's Brightest Bird.* Boston, New York: Houghton Mifflin Harcourt, 2016.

Websites

Frequently Asked Questions About Crows
birds.cornell.edu/crows/crowfaq.htm

Seven Simple Actions to Help Birds
birds.cornell.edu/home/seven-simple-actions-to-help-birds/

Corvid Research
corvidresearch.blog

Winter Crow Roost
wintercrowroost.com/crows-in-the-news/

All About Birds
allaboutbirds.org

Merlin Bird ID
merlin.allaboutbirds.org

eBird
ebird.org

iNaturalist
inaturalist.org

Index

A

aerial super-sliding, 53
age, 74
alarm calls, 59
altricial, 8
American Crows (*Corvus brachyrhynchos*), 2, 6, 16, 61, 73, 74, 80, 87. *See also* crows
American Robins, 27. *See also* robins
apps, naturalist, 8
Australian Ravens (*Corvus coronoides*), 7

B

banding birds, 31-32, 34
BIRDS
    banding (tagging), 31-32, 34
    bird-watching, 27
    brains, 38-39, 40, 47
    declining bird populations, 26-27
    evolution of, 10-11
    songbirds, 16, 26-27, 37, 54, 58
bird-watching, 27
body language, 59
bombs away, 53
brains, 38-39, 40, 47
breeding, cooperative, 16-17
budding, 89

C

caring, for sick and injured crows, 45, 81, 86
Carrion Crows (*Corvus corone*), 6
cats, 27
cawing, 54, 55, 56-57, 58-59, 89
Common Ravens (*Corvus corax*), 6
communication, 54, 55, 56-57, 58-59, 89
convergent evolution, 14
cooperative breeding, 16-17
Cooper's Hawk, 21, 27
coos, 59
corvids, 2
Corvus (genus), 2, 3
Corvus brachyrhynchos (American Crows), 2, 6, 16, 61, 73, 74, 80, 87. *See also* crows
Corvus corax (Common Ravens), 6
Corvus cornix (Hooded Crows), 6
Corvus corone (Carrion Crows), 6
Corvus coronoides (Australian Ravens), 7
Corvus frugilegus (Rooks), 7
Corvus hawaiiensis (Hawaiian Crow), 87
Corvus macrorhynchos (Large-billed Crows, Jungle Crows), 7
Corvus moneduloides (New Caledonian Crows), 7, 42
Corvus ossifragus (Fish Crows), 6
Corvus splendens (House Crows), 7
counting crows, 17
courtship, 67, 91
CROWS
    age, 74
    banding (tagging), 31-32, 34
    brains, 38-39, 40, 47
    caring for sick and injured crows, 45, 81, 86
    cawing, 54, 55, 56-57, 58-59, 89
    counting crows, 17
    courtship, 67, 91
    culture and, 32
    death and funerals, 76
    distance travelled in a day, 64
    eggs, 4, 20, 22, 94
    evolution of, 10-11
    extraordinary lives of, 92-93
    family dynamics, 16-17, 66-67, 68-69, 74-75, 85-86, 88-91
    fledglings, 33, 40, 44, 45, 46-47, 49-50
    food, 13-14, 25, 26-27, 28-29, 46-47, 64-65
    four-toed feet, 11

geographic range, 2, 6-7
hatchlings, 4-5, 8-9, 94
identifying, 3, 8, 33
intelligence, 38-39, 47, 89
maturity, 88-91, 94-95
migration, 61, 66
nestlings, 13-14, 16-18, 25-26, 30-31, 34, 37-38, 40
nests, 2-4, 15, 19, 22-23, 66, 94
physical differences by age, 33
play, 40, 42, 50-52, 53
poetry about, 47, 71
predators of, 21, 37, 59
protecting crows, 87
recognizing humans, 32
sexual maturity, 73-74
species of, 2, 6-7
stuck on the ground, 45
territory of, 18-19, 64-65
tool use, 42-43
West Nile virus and, 76-77, 79, 80
winter roost, 62-63, 66
yearlings, 67, 70, 74-75, 85-86
culture, 32

D
Darwin, Charles, 51
death, 76
dinosaurs, 11, 38
drop-and-catch, 40, 53

E
eggs, 4, 20, 22, 94
egg tooth, 5
EVOLUTION
    birds, 10-11
    convergent evolution, 14

F
family dynamics, 16-17, 66-67, 68-69, 74-75, 85-86, 88-91
feathers, 30
fecal sacs, 26
feet, four-toed, 11
Fish Crows (*Corvus ossifragus*), 6
fledglings, 33, 40, 44, 45, 46-47, 49-50
food, 13-14, 25, 26-27, 28-29, 46-47, 64-65
funerals, 76

G
Game of Crows, 68-69
games, 40, 42, 50-52, 53
Great Horned Owl, 21, 61, 85

H
haiku, 71
hatchlings, 4-5, 8-9, 94
Hawaiian Crow (*Corvus hawaiiensis*), 87
Hawk, Cooper's, 21, 27
Hooded Crows (*Corvus cornix*), 6

House Crows (*Corvus splendens*), 7
HUMANS
    banding (tagging) birds, 31-32, 34
    caring for sick and injured crows, 45, 81, 86
    finding your flock, 95
    protecting crows, 87
    recognized by crows, 32
    West Nile virus and, 78-79, 82-83

I
iNaturalist (app), 8
intelligence, 38-39, 47, 89. *See also* play; tools

J
Jungle Crows (Large-billed Crows, *Corvus macrorhynchos*), 7

L
Large-billed Crows (Jungle Crows, *Corvus macrorhynchos*), 7
larynx, 54
limerick, 47

M
maps, of known world, 65
Merlin Bird ID (app), 8
migration, 61, 66
Migratory Birds Convention, 87
mob, 21, 32
mosquitos, 77, 78, 82-83
murder, of crows, 30

N

nestlings, 13-14, 16-18, 25-26, 30-31, 34, 37-38, 40
nests, 2-3, 15, 19, 22-23, 66, 94
New Caledonian Crows (*Corvus moneduloides*), 7, 42

O

Owl, Great Horned, 21, 61, 85

P

people. *See* humans
pip, 4
play, 40, 42, 50-52, 53
POETRY
    haiku, 71
    limerick, 47
poop, 26
precocial, 8
predators, 21, 37, 59

R

race-and-chase, 53
rattle calls, 59
RAVENS
    Australian Ravens (*Corvus coronoides*), 7
    Common Ravens (*Corvus corax*), 6
robins, 27, 37, 66
Rooks (*Corvus frugilegus*), 7
roost, winter, 62-63, 66

S

sentinel species, 79
sexual maturity, 73-74
Shakescrow, William, 91
snow sports, 53
social dynamics, 16-17, 66-67, 68-69, 74-75, 85-86, 88-91
songbirds, 16, 26-27, 37, 54, 58
sounds, made by crows, 54, 55, 56-57, 58-59, 89
spectrograms, 55
spring, 66
squalling calls, 59
stay-at-home behavior, 74
string games, 41
structured caws, 59
syrinx, 54, 56

T

tagging birds, 31-32, 34
territory, 18-19, 64-65
therapods, 11
tools, 42-43
tug-of-war, 50, 53

U

umbilical scars, 9

V

virus, West Nile, 76-77, 78-79, 80, 82-83, 87
voice, 54. *See also* cawing

W

wanna fight?, 53
West Nile virus, 76-77, 78-79, 80, 82-83, 87
wildlife rehabilitation, 45, 81
winter, 61
winter roost, 62-63, 66

Y

yearlings, 67, 70, 74-75, 85-86

Z

zoonotic disease, 78

About the Author

Candace Savage is the award-winning author of more than two dozen books, many of which reflect her love of the living world. Her writing for young people has been honored by the Canadian Children's Book Centre and the New York Public Library, among others. In 2022, she received both the Cheryl and Henry Kloppenburg Award for Literary Excellence and the Matt Cohen Award: In Celebration of a Writing Life. She is privileged to live and write in Treaty Six territory and the homeland of the Métis Nation in Saskatchewan, Canada.

About the Illustrator

Rachel Hudson creates artwork that reconnects people with the natural world. She is the author and illustrator of 100 Endangered Species (Button Books, 2021).

Her research for this book included visiting an abandoned crow's nest in a climbing harness to see the world of a crow from the top of a pine tree.

When not in her studio in Hampshire, England, she's happiest marking the seasons and taking delight in surprise encounters with wildlife, however fleeting.

The David Suzuki Institute is a companion organization to the David Suzuki Foundation, with a focus on promoting and publishing on important environmental issues in partnership with Greystone Books.

We invite you to support the activities of the Institute. For more information, please contact us at:

David Suzuki Institute
219 – 2211 West 4th Avenue
Vancouver, BC, Canada V6K 4S2
info@davidsuzukiinstitute.org
604-742-2899
davidsuzukiinstitute.org

Cheques can be made payable to The David Suzuki Institute.

Text copyright © 2024 by CANDACE SAVAGE
Illustrations copyright © 2024 by RACHEL HUDSON

24 25 26 27 28   5 4 3 2 1

All rights reserved. No part of this book may be reproduced, stored in a retrieval system or transmitted, in any form or by any means, without the prior written consent of the publisher or a license from The Canadian Copyright Licensing Agency (Access Copyright). For a copyright license, visit accesscopyright.ca or call toll free to 1-800-893-5777.

Greystone Kids / Greystone Books Ltd.
greystonebooks.com

David Suzuki Institute
davidsuzukiinstitute.org

Cataloguing data available from Library and Archives Canada
ISBN 978-1-77164-916-2 (cloth)
ISBN 978-1-77164-917-9 (epub)

Editing by Linda Pruessen
Copy editing by Dawn Loewen
Proofreading by Tracy Bordian
Cover and interior design by Jessica Sullivan
The illustrations in this book were rendered in mixed media, including hand-printed textures, brought together digitally.

Printed and bound in China on FSC® certified paper at Shenzhen Reliance Printing. The FSC® label means that materials used for the product have been responsibly sourced.

Greystone Books thanks the Canada Council for the Arts, the British Columbia Arts Council, the Province of British Columbia through the Book Publishing Tax Credit, and the Government of Canada for supporting our publishing activities.

Greystone Books gratefully acknowledges the xʷməθkʷəy̓əm (Musqueam), Sḵwx̱wú7mesh (Squamish), and səlilwətaɬ (Tsleil-Waututh) peoples on whose land our Vancouver head office is located.